T0260404

WORK DOMAIN ANALYSIS

ANALYSIS

CONCEPTS, GUIDELINES, AND CASES

WORK DOMAIN ANALYSIS

ANALYSIS

CONCEPTS, GUIDELINES, AND CASES

Neelam Naikar

CRC Press
Taylor & Francis Group
Boca Raton   London   New York

CRC Press is an imprint of the
Taylor & Francis Group, an **informa** business

CRC Press
Taylor & Francis Group
6000 Broken Sound Parkway NW, Suite 300
Boca Raton, FL 33487-2742

© 2013 by Taylor & Francis Group, LLC
CRC Press is an imprint of Taylor & Francis Group, an Informa business

No claim to original U.S. Government works

Printed on acid-free paper
Version Date: 20130315

International Standard Book Number-13: 978-0-8058-6129-7 (Hardback)

Library of Congress Cataloging-in-Publication Data

Naikar, Neelam, 1969-
  Work domain analysis : concepts, guidelines, and cases / Neelam Naikar.
  pages cm
  Includes bibliographical references and index.
  ISBN 978-0-8058-6129-7 (hardback)
  1. Job analysis. 2. Task analysis. I. Title.

HF5549.5.J6N35 2013
658.3'06--dc23   2012020737

Visit the Taylor & Francis Web site at
http://www.taylorandfrancis.com

and the CRC Press Web site at
http://www.crcpress.com

This book is dedicated to Jens Rasmussen and Kim Vicente.

The more I delved into their work, the more awed I became.

Contents

Section III   Guidelines

Section V   Appendix

Preface

There is no question that workers in complex sociotechnical systems—such as hospitals, military aircraft, nuclear power plants, and emergency management organizations—have extraordinarily difficult jobs. It is also undeniable that to support people in these challenging roles we must first understand the nature of their work. Without a keen appreciation of their work, it is not possible to create designs that will enable people to meet the demands of their jobs successfully—in a way that is safe, productive, and healthy (Vicente, 1999). Instead, the designs of interfaces, teams, or instructional systems, for example, are likely to be inadequate.

Work domain analysis is the first of five dimensions of cognitive work analysis, a framework for the analysis, design, and evaluation of work in complex sociotechnical systems. Unlike other techniques for work analysis, this framework defines a system's work demands in terms of the constraints on actors' behavior rather than the behavior itself. By focusing on constraints, cognitive work analysis leads to designs that support flexibility or adaptation in the workplace instead of brittle designs that promote fixed patterns of behavior. The five dimensions of this framework focus on different classes of constraints.

Work domain analysis is concerned with the constraints that are placed on actors by the *work domain,* or the functional structure of the physical, social, or cultural environment in which they work. These constraints, which include a system's purposes and physical resources, are the fundamental boundaries on actors' behavior as they delineate the full set of possibilities for action at any moment. Designs that take into account such constraints can support actors in creating innovative patterns of behavior, which is essential, not only for productivity, but also for dealing with unforeseen events—situations that pose immense threats to a system's viability. Moreover, such designs can support actors in following their individual preferences when it is appropriate, which promotes a healthier workforce.

The objective of this book is to make work domain analysis more accessible to readers. I do not aim to do this by oversimplifying the principles of this form of analysis but, instead, by explaining its concepts and methodology in considerable depth. As well as attending to fundamental topics such as abstraction, decomposition, and structural means–ends and part–whole relations—the basic characteristics of a work domain model—I deal with more advanced topics such as the development of multiple models of a system and the distinction between causal and intentional systems. These topics are illustrated with numerous examples, crossing a range of systems, including warships, libraries, and petrochemical plants. One of the systems I refer to, a home, will be highly familiar to readers, which means that their comprehension of work domain analysis will not be limited by their knowledge of this system. I believe a firm grasp of work domain analysis will lead to more powerful and innovative applications of this approach. It will also make work domain analysis easier and more enjoyable for analysts to perform.

Background

This book builds on the foundational texts on cognitive work analysis by Rasmussen (1986), Rasmussen, Pejtersen, and Goodstein (1994), and Vicente (1999). These texts provide an authoritative account of the conceptual foundations of this framework, including work domain analysis. A thorough grounding in these texts is essential for understanding cognitive work analysis and applying it well.

The motivation behind this book is the fact that work domain analysis remains relatively challenging for many readers. One reason why this technique may be challenging is that the earlier texts cover all five dimensions of cognitive work analysis, which means that the concepts of its first dimension alone cannot be treated at length. Another is that these texts focus on providing a conceptual perspective of the framework because it was still relatively unknown at the time (Vicente, 1999). As a result, extensive guidance for performing work domain analysis could not be provided. In addition, because of differences in terminology, these books appear to present somewhat different approaches to work domain analysis, which can be confusing for many readers. Finally, these texts illustrate the concepts of work domain analysis with fairly specialized systems, so that it can be difficult for readers to extract general principles that can be applied to new or different systems.

Purpose and Scope

This book makes work domain analysis more accessible to readers in three main ways. First, it contributes not only to the articulation but also to the development of this form of analysis. The concepts of work domain analysis are recounted in considerable depth, including topics that are not addressed in detail—or at all—in the earlier texts. In addition, for the first time, extensive guidelines for performing this type of analysis are provided.

Second, the concepts and guidelines are illustrated with examples from a home, a system that readers will recognize. Readers, therefore, should not find this material unduly challenging to comprehend, as may have been the case with less familiar examples. The home examples also complement the case study of DURESS II, a thermalhydraulic microworld simulation which Vicente (1999) uses to illustrate his book. Whereas DURESS II is a *causal system*, in which actors' behavior is constrained primarily by physical or natural laws, the examples provided here are from an *intentional system*, in which actors' behavior is constrained primarily by social laws, conventions, or values.

Third, three case studies are presented of the application of work domain analysis to large-scale systems in industrial settings. These case studies serve two aims: to provide additional examples of the guidelines for work domain analysis and to demonstrate that this type of analysis can be applied to a variety of problems in industrial settings, specifically to evaluate design concepts, to develop team designs, and to examine training needs and instructional system requirements. Although these case studies have been reported previously, the process of performing work domain analysis for these applications was not discussed at length. The impact, uniqueness, and feasibility of work domain analysis for such applications also were not assessed in detail. Without an appreciation for the variety of applications of work domain analysis and its suitability for implementation in industrial settings, the benefits of this approach may not be fully realized.

Many of the topics in this book reflect the challenges I faced in learning work domain analysis and applying it in industrial settings, as well as the questions inexperienced analysts have asked me about this technique. By highlighting apparent differences between the texts by Rasmussen (1986), Rasmussen et al. (1994), and Vicente (1999) in addressing some of these topics, it is not my intention to polarize those working in this area. Rather, I make these observations because the apparent differences between the texts can be a source of confusion for people using them.

The guidelines for work domain analysis presented in this book were informed by the concepts of this form of analysis, a broad range of case studies of its application, and evaluations of earlier versions of the guidelines. My aim in proposing guidelines is not to suggest that it is possible or desirable to prescribe a process for performing this type of analysis. On the contrary, I believe that it is both impossible and undesirable to be prescriptive in this context. That is why the guidelines are presented as a set of analytic themes or questions for analysts to consider when performing work domain analysis, not as a series of steps for analysts to execute. Inexperienced analysts may find these guidelines beneficial as a starting point, but they may develop variations or extensions that suit their requirements, or those of specific projects, as they become skillful.

The home examples may seem simplistic at times. As noted previously, a home was selected as the primary system for illustration with the intention of making work domain analysis easier for readers to understand. Often, much of the principal material is also illustrated with examples from other systems, such as libraries, hospitals, and emergency management organizations.* The three case studies, which describe the application of work domain analysis to military systems, provide more complex illustrations as well.

To emphasize that work domain analysis is only one of the dimensions of cognitive work analysis, the key concepts of the remaining dimensions are summarized in the Appendix. These dimensions—control task analysis, strategies analysis, social organization and cooperation analysis, and worker competencies analysis—are also illustrated with home examples. It is important to remember that not all of the information pertinent to the analysis, design, and evaluation of work in complex sociotechnical systems falls within the purview of work domain analysis; some of this information is addressed within the other dimensions.

Structure

This book has five sections. In the first section, which comprises Chapters 1 and 2, cognitive work analysis and work domain analysis are introduced. Chapter 1 provides an overview of the five dimensions of cognitive work analysis and explains the rationale for the order of those dimensions. How this framework differs from other approaches to work analysis and why it supports designing for adaptation are also addressed here.

Chapter 2 focuses on work domain analysis. It distinguishes the analysis of actors' *behavior* from that of the *environment* in which actors behave, and it explains why models

---

* Note that these models have been reproduced to match the original presentation as closely as possible, with only obvious errors corrected. In addition, many of the models contain abbreviations that have not been defined further in this book, mainly because this information is unneccessary for understanding the concepts that the models are intended to illustrate.

of the environment—or work domain—are useful. The abstraction–decomposition space and the abstraction hierarchy, the main modeling tools for work domain analysis, are described, and a model of a home is supplied for illustration. This chapter also explains the distinction between causal and intentional systems and discusses four formats that analysts commonly use to present work domain models: the graphical abstraction–decomposition space, the graphical abstraction hierarchy, the tabular abstraction–decomposition space, and the tabular abstraction hierarchy. These topics are important for understanding the succeeding material.

Section II, which has five chapters, examines the concepts of work domain analysis in more depth, focusing on topics that were not covered in sufficient detail in the earlier texts. These chapters are organized largely around the abstraction–decomposition space. Chapter 3 is concerned with the abstraction dimension of this modeling tool, whereas Chapter 4 considers its decomposition dimension. Chapter 5 focuses on the hierarchical relationships that define the abstraction–decomposition space, namely, structural means–ends and part–whole relations. It also introduces topological relations, which are sometimes included in a work domain model. Chapter 6 discusses two reasons why analysts have created multiple models of a system, which are to represent different stakeholders' perspectives of a problem and to highlight distinct aspects of a problem. Chapter 7 addresses the relationship between the analysis of activity and work domain analysis.

The third section of the book consists of a single chapter—Chapter 8—which presents guidelines for work domain analysis. As mentioned before, these guidelines are in the form of a set of analytic themes or questions rather than a series of steps. This format reflects the fact that performing work domain analysis is not a fixed and linear process, but a flexible and iterative one.

Section IV, which has three chapters, presents detailed case studies of the application of work domain analysis to large-scale systems in industrial settings. These applications include evaluating competing design concepts (Chapter 9); designing teams for future, first-of-a-kind systems (Chapter 10); and examining training needs and instructional system requirements (Chapter 11).

The final section comprises an Appendix that summarizes the key concepts of the remaining four dimensions of cognitive work analysis.

Audience

This book is intended for students, researchers, educators, and practitioners who would value a comprehensive treatment of work domain analysis. As stated previously, I believe a sound appreciation of this form of analysis will allow analysts to apply it more powerfully and imaginatively. It will also make work domain analysis easier and more enjoyable for analysts to perform.

While the treatment of topics in this book is fairly detailed, I have made every effort to make it easy to read and understand. Finding the right balance in writing style to suit both novices and those who are more familiar with work domain analysis has been challenging. Most often, I have chosen to write for readers with less experience. This means that when information is presented in a way that seems pedantic or tedious, I have most likely done this deliberately for the sake of clear instruction. I believe this presentation may be helpful when beginners, especially those without access to experienced analysts, utilize this material to build a work domain model. That said, it is not my view that novices can become proficient in work domain analysis solely from this book. The texts by Rasmussen et al. (1994) and Vicente (1999), in particular, are critical, and advice from experts is beneficial.

Terminology

Terms relating to cognitive work analysis are, for the most part, used consistently with a glossary provided by Vicente (1999). The meanings of these terms are conveyed at suitable levels of detail at appropriate points in this text and are searchable through the index. Those terms I use differently from Vicente are described explicitly here, although the full meaning of these terms is perhaps best understood within the context of their use. My and Vicente's usages of these terms are not fundamentally different from each other. My particular usage reflects the specific needs of this book.

I use the term *system* to refer only to a sociotechnical system; Vicente (1999) uses this term to mean a work domain or actor (i.e., worker or automation) as well. I adopt his definition of *sociotechnical system*, which is "A System composed of technical, psychological, and social elements" (p. 9). I also accept his definition of *complex sociotechnical systems*, which is "Sociotechnical Systems that rate highly on several of the following dimensions: large problem space, social, heterogeneous perspectives, distributed, dynamic, potentially high hazards, many coupled subsystems, automated, uncertain data, mediated interaction via computers, disturbance management" (p. 5). Many of the sociotechnical systems I discuss in this book are complex sociotechnical systems.

By *work domain*, I mean the functional structure of the physical, social, or cultural environment of actors in a system, which places constraints on their behavior. Work domain analysis, therefore, is an analysis of this functional structure. Like Vicente (1999), I use *functional* to mean "action-relevant" (p. 155), or applicable for performing action, and *structure* to refer to "A relatively permanent relational property of a System" (p. 9). Hence, work domain analysis targets the action-relevant, relatively permanent relational properties of actors' environments. Whenever I mention *structural properties*, I mean action-relevant structural properties, although I do not say so explicitly each time.

Vicente (1999) defines *work domain* as "The System being controlled, independent of any particular Worker, Automation, Event, Task, Goal, or Interface" (p. 10). This definition may be taken to mean that an analysis or model of a work domain cannot incorporate control systems, including humans and automation. As discussed in Chapter 7, however, it is possible to represent the structural properties of control systems in a work domain model, although not their actions or behavior. Thus, to avoid confusion, I define *work domain* in a way that differs from Vicente's, although this description is not fundamentally different from his general usage of the term.

Finally, I use the term *subject matter experts*, which is not defined by Vicente (1999), to mean people whom analysts may consult as sources of information for an analysis. For work domain analysis, these are people who have knowledge of the system under study. Workers, then, are key subject matter experts, but a range of other people is usually suitable as well. For instance, when conducting an analysis of a home, the subject matter experts may include the inhabitants as well as appropriate tradespeople, visitors, builders, and architects.

Acknowledgments

First, I would like to convey my heartfelt gratitude to the Defence Science and Technology Organisation for the tremendous support it provided to me in writing this book. Particular thanks are due to past and present members of the Centre for Cognitive Work and Safety Analysis for their important contributions to our research program, as reflected in this book. I am also grateful to a number of senior managers for their unwavering support over the writing period: James Meehan, Torgny Josefsson, David Graham, and Simon Parker.

Special mention is owed to numerous subject matter experts and personnel from the Royal Australian Air Force, and other parts of the Australian Department of Defence, for the significant part they played in our applications of cognitive work analysis to a range of defense problems. I am deeply grateful not only for their generosity in providing my colleagues and me with insight into the details of their work practices, and sharing their knowledge of defense operations, but also for their openness in considering the use of novel approaches on their projects. In particular, I would like to acknowledge Ms. Tracey Bryan, Mr. Ben Hall, Group Captain Antony Martin, Squadron Leader Paul Nowland, Air Commodore Christopher Westwood, and Squadron Leader Carl Zell.

Several colleagues helped me in immeasurable ways in the writing and preparation of this book for publication, especially by providing me with valuable feedback on a number of earlier drafts. For this, I am indebted to Kate Branford from Dédale Asia Pacific and Ashleigh Brady, Ben Elix, Sandra Lambeth, Russell Martin, James Meehan, Elissa Scuderi, Alanna Treadwell, and Jenny Yeung from the Defence Science and Technology Organisation. My sincere thanks also go to Greg Jamieson from the University of Toronto and Cathy Burns from the University of Waterloo for their comments on an earlier draft. In addition, I am grateful to Jan O'Reilly, the publications manager at the Defence Science and Technology Organisation, for her advice and Adam Woollett from Boeing Australia Limited for his assistance in creating the figures for this book.

Thanks also to my editors, initially Anne Duffy from Lawrence Erlbaum Associates and later Cindy Renee Carelli, senior acquisitions editor, and Amy Blalock, senior project coordinator from Taylor & Francis, for their assistance, especially in dealing with my inquiries promptly and patiently. A special thank you as well to my project editor, Marsha Hecht, and her team for their help during the final production process.

Finally, my deep appreciation to my family—Russell Martin, and Bhakta, Kamala, and Mithran Naiker—for the time and space they allowed me for my work.

About the Author

Neelam Naikar leads the Centre for Cognitive Work and Safety Analysis at the Defence Science and Technology Organisation in Melbourne, Australia. Following her academic training in Psychology, she was drawn to studying human behavior in complex engineered systems. Her interest in cognitive work analysis—of which work domain analysis is an important dimension—was captured when she recognized its potential for modeling the cognitive demands on workers in complex sociotechnical systems in a form that could provide a powerful approach to system design. For the 15 years before this publication, Dr. Naikar has conducted extensive research into the cognitive work analysis framework: its concepts, methodology, and applications. Her major scientific contributions include the development of methods for performing cognitive work analysis and the extension of the application of this framework beyond interface design to a range of other problems, such as the evaluation of system design concepts, team design, and training. Her research is informed not only by a scholarly knowledge of a range of work analysis techniques but also by the practical application of these techniques in industry. Dr. Naikar graduated with a Bachelor of Science degree with Honors in Psychology from the University of New South Wales, Australia, in 1993, and a PhD in Psychology from the University of Auckland, New Zealand, in 1996. She is currently a member of the editorial boards of the *International Journal of Aviation Psychology* and the *Journal of Cognitive Engineering and Decision Making.*

Section I

Introduction

1

Cognitive Work Analysis

OVERVIEW Cognitive work analysis is a framework for the analysis, design, and evaluation of work in complex sociotechnical systems. In this chapter, the five dimensions of this framework—work domain analysis, control task analysis, strategies analysis, social organization and cooperation analysis, and worker competencies analysis—are introduced. The significance attached to the ordering of these dimensions is also revealed. A case is made that this framework for work analysis is unique because of its formative quality. Moreover, it is explained that this framework is essential for developing highly effective systems as it promotes designing for adaptation. This overview of cognitive work analysis provides sufficient context for the following chapters on work domain analysis, the first dimension of this framework. A more detailed discussion of the other dimensions may be found in the Appendix.

Work Analysis

Few would doubt that workers in complex sociotechnical systems have highly challenging jobs. Consider, for instance, hospitals, nuclear power plants, fighter aircraft, warships, and emergency management organizations. Experts in these systems must possess deep knowledge of complex physical or natural processes, such as the aerodynamic processes of flight or the biological processes of the human body. Not only that, their roles have a strong social or cultural dimension as well because they must coordinate their activities with coworkers or provide assistance to members of the public, for example. Workers must weigh what is feasible technically against what is desirable socially or culturally. Trade-offs may be necessary between conflicting priorities, frequently under stressful conditions, including time pressure. Errors may be costly, not only financially or materially, but also to human life.

To support workers in these challenging roles, we first need to understand the nature of their work. Among other things, we need to appreciate the purposes they must fulfill, the priorities they have to balance, the activities they must perform, the decisions they have to make, the strategies they can adopt, and the equipment they can use. Only then is it possible to produce designs for workers that will enable them to meet their work demands successfully—in a way that is safe, productive, and healthy (Vicente, 1999). Otherwise, the designs of interfaces, teams, or instructional systems, for instance, are likely to fall short of the mark. Successful design depends on true knowledge of a system's work demands, even if that information is only understood implicitly, rather than analyzed systematically.

There are countless techniques for work analysis, too numerous to list here. Hierarchical task analysis (e.g., Shepherd, 2001), cognitive task analysis (e.g., Schraagen, Chipman, and Shalin, 2000), and cognitive work analysis (Rasmussen, 1986; Rasmussen, Pejtersen, and Goodstein, 1994; Vicente, 1999) are some well-known ones. These techniques define a system's work demands in radically different ways. They reflect diverse views of human behavior, or work practice, and they have unique design philosophies. This means that they make particular assumptions, not only about the nature of human work but also about how this work is ideally supported. The question of which technique to adopt, therefore, requires careful consideration. This book is centered on cognitive work analysis, a framework that is gaining increasing attention from researchers and practitioners. The benefits of this approach, as well as its distinctiveness from other techniques, are made clear in this chapter.

What Is Cognitive Work Analysis?

Cognitive work analysis is a framework that defines the work demands of complex socio-technical systems in terms of the constraints on actors (Rasmussen, 1986; Rasmussen et al., 1994; Vicente, 1999). Many different disciplines use the term *constraints*. For example, this term features in the language of mathematicians and engineers. However, this term has different meanings in different contexts. In the case of cognitive work analysis, constraints are limits on behavior, which must be respected by actors for a system to perform effectively (Figure 1.1). Constraints distinguish behaviors that are possible or acceptable from those that are impossible or unacceptable. Cognitive work analysis, therefore, is concerned with constraints that are "behavior-shaping" (Rasmussen et al., 1994, p. 25). These constraints remain relatively constant or invariant across a range of circumstances, which means that they must be respected by actors in an array of situations.

Although constraints place limits on behavior, they still afford actors many degrees of freedom for action, as indicated by the trajectories (arrows) in the center of Figure 1.1. For instance, there are many different sequences in which actors can execute tasks, or many different strategies that actors can use for decision making, while remaining within the

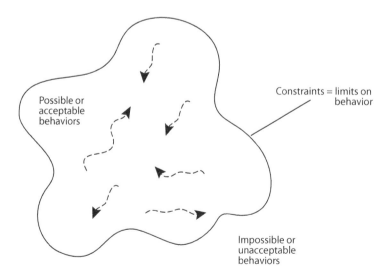

FIGURE 1.1
Constraints place limits on behavior but still afford actors many possibilities for action.

boundaries of effective performance. So, although constraints place limits on action, they usually do not uniquely specify the behavior of actors (Vicente, 1999).

To offer a simple illustration, a system's purposes and physical resources place constraints on actors' behavior (Figure 1.2). It will always be unacceptable to engage in behaviors that are inconsistent with a system's purposes, and it will always be impossible to use physical resources to perform behaviors they do not have the functionality to support. Nevertheless, within these constraints, actors have many possibilities for behavior that are compatible with a system's purposes and physical resources.

Complex sociotechnical systems can impose several kinds of constraints on actors. Accordingly, as shown in Table 1.1, cognitive work analysis comprises a number of dimensions of analysis, which focus on different types of constraints. The first two columns of the table list these dimensions as they are labeled by Vicente (1999) and Rasmussen et al. (1994). The next column notes the types of constraints with which each dimension is concerned. (Terms from both Vicente and Rasmussen et al. are used for describing these constraints.) The last column identifies the modeling tools for each dimension. In this book, I adopt Vicente's labels for the dimensions of the framework because of their prevalence in the recent literature.

Figure 1.3 shows that the dimensions of cognitive work analysis collectively define a constraint-based space within which actors have many possibilities for behavior. By focusing on constraints, cognitive work analysis promotes *designing for adaptation*. This means that, rather than developing designs that support workers in particular ways of working, which have been prescribed or described by analysts a priori, the aim is to create designs that support workers in adapting their behavior as they see fit without violating the system's constraints. With such designs, workers can create innovative patterns of behavior, which may be particularly important for dealing with unforeseen events, or they can adopt ways of working that suit their individual preferences, while all the time remaining within the boundaries of effective performance. This approach to design is essential for complex sociotechnical systems, as explained in more detail later in this chapter.

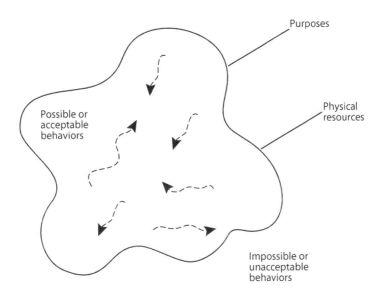

FIGURE 1.2
Within the constraints of a system's purposes and physical resources, actors have many possibilities for action.

TABLE 1.1

Dimensions of Cognitive Work Analysis

Vicente (1999)	Rasmussen et al. (1994)	Types of Constraints	Modeling Tools
Work domain analysis	Work domain analysis	Physical, social, or cultural environment	Abstraction–decomposition space, abstraction hierarchy
Control task analysis	Activity analysis in work domain terms and activity analysis in decision making terms	Work situations, work functions, or control tasks	Contextual activity template, decision ladder template
Strategies analysis	Activity analysis in terms of mental strategies	Strategies	Information flow map
Social organization and cooperation analysis	Analysis of the work organization	Allocation, distribution, or coordination of work	Abstraction–decomposition space, abstraction hierarchy, contextual activity template, decision ladder template, information flow map
Worker competencies analysis	Analysis of system users	Worker competencies	Skills, rules, and knowledge taxonomy

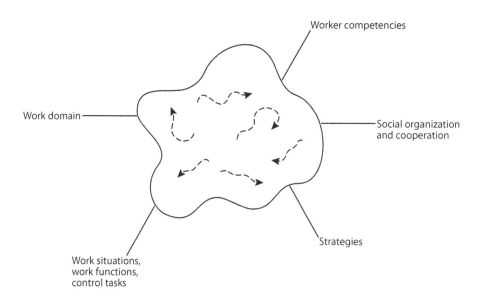

FIGURE 1.3

The dimensions of cognitive work analysis collectively define a constraint-based space within which actors have many degrees of freedom for action.

In what follows, I provide a further—but succinct and simplified—description of the dimensions of cognitive work analysis, with the intention of imparting a little more information about the constraints associated with them. A few examples from a familiar system, a home, are supplied for each dimension. Work domain analysis is then discussed at length in subsequent chapters. A more detailed treatment of the remaining dimensions is provided in the Appendix.

Work Domain Analysis

Work domain analysis focuses on the constraints associated with the functional structure of the physical, social, or cultural environment of actors. The structural properties of actors' environments define the fundamental *reasons* and *resources* for their behavior. These properties include a system's purposes as well as its physical objects. The purposes of a home, for example, may include well-being, and its physical objects may include furniture. The functional structure of the environment, then, places constraints on actors' behavior by specifying the purposes to be achieved with the physical objects that are available. The main modeling tools for this dimension are the abstraction–decomposition space (Figure 1.4a) and the abstraction hierarchy (Figure 1.4b).

Control Task Analysis

Control task analysis is concerned with the constraints associated with *what* needs to be done in a system. It identifies the activity that is required for achieving a system's purposes with a suite of physical resources. This activity may be characterized as a set of recurring *work situations*, *work functions*, or *control tasks* (Naikar, Moylan, and Pearce, 2006; Rasmussen et al., 1994; Vicente, 1999). Consider the following examples for a home: the work situations for actors to participate in may be different stages of the day, such as morning and evening; the work functions to be performed may include preparing meals and beverages; and the control tasks for achieving this work function may involve decisions about which meals or beverages to prepare. The contextual activity template (Figure 1.5a) provides a tool for modeling work situations and work functions, whereas the decision ladder template (Figure 1.5b) can be used for modeling control tasks.

Strategies Analysis

Strategies analysis focuses on the constraints associated with the different ways in which activity can be accomplished. Whereas control task analysis is concerned with *what* activity is needed, strategies analysis is concerned with *how* the activity can be done. Several strategies are usually possible for a single activity. In the case of a home, it is possible to prepare a meal by following a recipe, by relying on one's memory, or by trial and error. Strategies analysis, therefore, involves examining the range of strategies that are possible for an activity. A strategy may be portrayed graphically with an information flow map, but this modeling tool is not yet available in a generic format. So far, information flow maps have only been created for particular applications (Figure 1.6).

Social Organization and Cooperation Analysis

Social organization and cooperation analysis is concerned with the constraints that the *allocation*, *distribution*, and *coordination* of work impose on actors. Typically, there are many

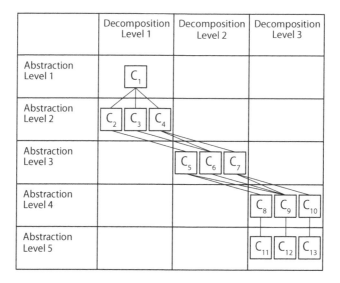

(a)

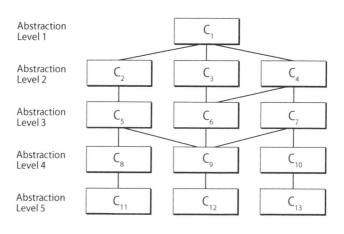

(b)

FIGURE 1.4
The main modeling tools for work domain analysis: (a) abstraction–decomposition space; (b) abstraction hierarchy. C, constraint. More information about these tools is provided in Chapter 2.

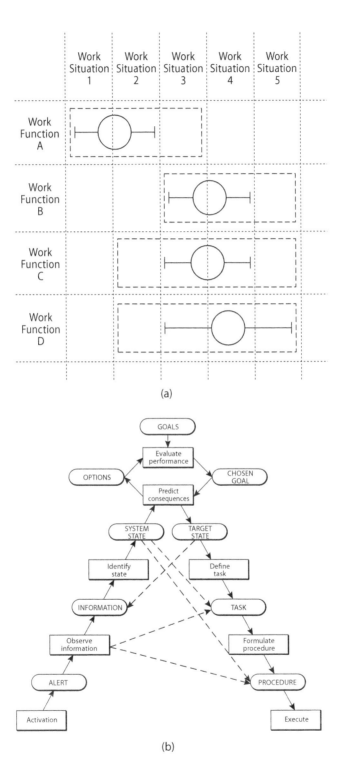

(a)

(b)

FIGURE 1.5
The main modeling tools for control task analysis: (a) contextual activity template; (b) decision ladder template. These templates are explained further in the Appendix.

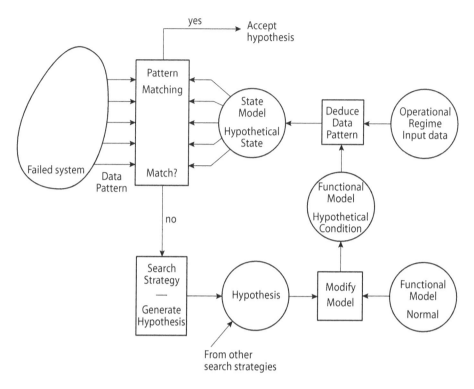

FIGURE 1.6

An information flow map, the main modeling tool for strategies analysis, representing a hypothesis and test search strategy employed by professional technicians in repairing electronic equipment. This strategy is explained further by Rasmussen (1981) and Vicente (1999). (Adapted from Rasmussen, J., 1981, Models of mental strategies in process plant diagnosis, In J. Rasmussen and W. B. Rouse, Eds., *Human detection and diagnosis of system failures,* Plenum Press, New York, 241–258. © 1981 Plenum Press (Figure 4). With permission of Springer Science + Business Media.)

possibilities for work organization in a system. These possibilities may be specified in relation to the work domain, work situations, work functions, control tasks, or strategies. For instance, in a home, the organizational structure may be specified by work functions, so that one inhabitant is responsible for preparing meals and another for conducting housework. The modeling tools from the preceding dimensions are useful for examining the possible ways that work can be organized in a system. Figure 1.7 illustrates how the decision ladder template may be utilized for this purpose.

Worker Competencies Analysis

Worker competencies analysis focuses on the constraints associated with the competencies necessary for meeting a system's work demands. The preceding dimensions specify work demands, which collectively shape the competencies workers need. For example, if certain strategies for preparing a meal in a home (strategies analysis) are always executed by automated devices (social organization and cooperation analysis), inhabitants will require a different set of competencies from those necessary for executing those strategies themselves. Human cognitive capabilities and limitations place constraints on how a system's work demands can be met. The skills, rules, and knowledge taxonomy (Figure 1.8)

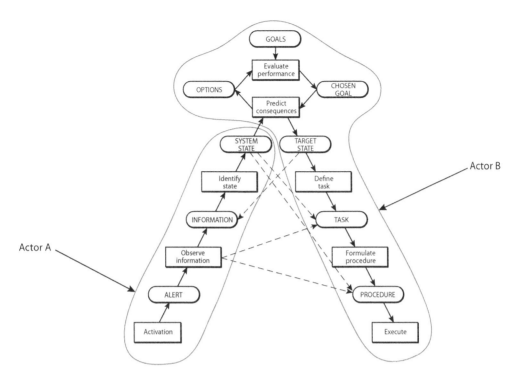

FIGURE 1.7
The decision ladder template provides a tool for social organization and cooperation analysis; this figure shows one possibility for the distribution of control tasks across two actors. The Appendix has more information about the use of the decision ladder template for this dimension.

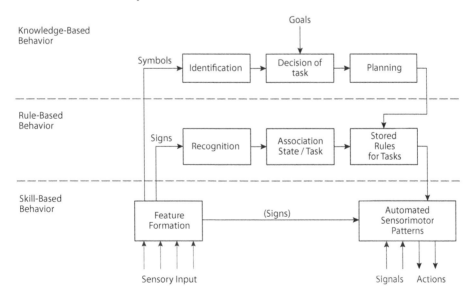

FIGURE 1.8
The skills, rules, and knowledge taxonomy, the main tool for worker competencies analysis. This taxonomy is described in more detail in the Appendix. (Adapted from Rasmussen, J., 1983, Skills, rules, and knowledge; signals, signs, and symbols, and other distinctions in human performance models, *IEEE Transactions on Systems, Man, and Cybernetics, 13*(3), 257–266. Copyright © 1983 IEEE. With permission.)

provides a framework for integrating information about work demands with knowledge about human cognition in a way that makes the implications for design clearer.

Order of the Five Dimensions

The five dimensions of cognitive work analysis are not ordered as methodological steps that *must* be followed in sequence. Instead, the ordering reflects a number of conceptual relationships, signifying the ecological orientation of this framework (Vicente, 1999). Some of these relationships are described here.

The order of the five dimensions is such that they start with the constraints of the work domain, or environment, and end with those of the cognitive characteristics of workers. Each successive dimension represents a step away from environmental constraints, which are those that originate in the physical, social, or cultural context in which workers are situated, and a step toward cognitive constraints, which are those that originate in the human cognitive system. This ordering reflects the precedence cognitive work analysis gives to environmental constraints over cognitive constraints—or its ecological orientation.

An ecological approach to work analysis emphasizes that designs must be compatible with environmental constraints before it is worth ensuring that they are compatible with cognitive constraints. Compare this with a cognitivist approach, which advocates that designs must be congruous with workers' mental models. From an ecological perspective, the cognitivist approach is pointless if workers' mental models do not correspond with reality, that is, with the reality of their environment. This does not mean that cognitive constraints are of no consequence to cognitive work analysis. Rather, this framework is concerned with both environmental and cognitive constraints but gives precedence to the former.

The order of the five dimensions is also such that options or possibilities for behavior are reduced across successive dimensions. Figure 1.9 is an alternative depiction of the constraint-based space of the five dimensions to that shown in Figure 1.3. This figure shows that work domain constraints delineate the total set of behaviors that are possible or acceptable at any moment. These constraints, which include a system's purposes and physical objects, remain relevant regardless of the activities pursued, the strategies adopted, the organizational structures established, or the competencies utilized. However, the set of behavioral possibilities becomes progressively smaller depending on the activities, strategies, organizational structures, or competencies that are invoked.

More specifically, once actors have chosen an activity (e.g., Activity 2 in Figure 1.9), the possibilities for behavior are smaller than the total set delineated by the work domain. The subset of possibilities is reduced even further when a strategy is selected for an activity (e.g., Strategy A in Figure 1.9). After an organizational structure (e.g., one worker plus automation or two workers) is established for executing a strategy, the possibilities decrease yet again. This subset becomes smaller once certain competencies (e.g., memory, knowledge, or computational effort) are necessitated by the organizational structure. Nevertheless, within that subset, actors still have many possibilities for behavior.

The behavioral possibilities at any instant are dynamic, varying according to the combination of activities, strategies, organizational structures, and competencies that is active.

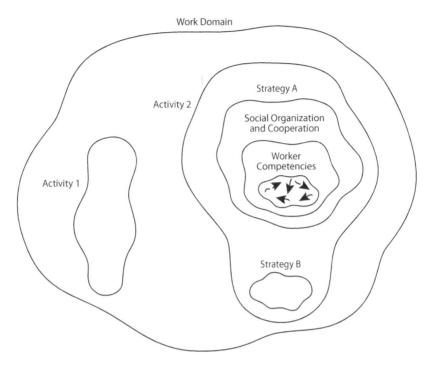

FIGURE 1.9
Each successive dimension of cognitive work analysis represents a reduction in the possibilities for action. (Adapted from Vicente, K. J., 1999, *Cognitive work analysis: Toward safe, productive, and healthy computer-based work*, Erlbaum, Mahwah, NJ. With permission of Taylor and Francis Group LLC. Permission conveyed through Copyright Clearance Center, Inc.)

However, as mentioned earlier, the behaviors that are possible or acceptable at any time can never exceed the set defined by the work domain. In any system, it will always be unacceptable to engage in behaviors that are inconsistent with its purposes, and it will always be impossible to use its physical objects to perform behaviors they do not have the functionality to support. Work domain analysis, therefore, defines the "fundamental bedrock" of constraints on actors (Vicente, 1999, p. 115).

Finally, the order of the five dimensions reflects the logical precedence of decisions with which analysts or designers are confronted. Before deciding what activities are necessary in a system (control task analysis), it is worth knowing the functional structure of the environment of actors (work domain analysis), including such things as the system's purposes and physical objects. Similarly, prior to determining what strategies are possible for an activity (strategies analysis), it is useful to understand what activities are necessary in the first place (control task analysis). Decisions about how work can be allocated among actors (social organization and cooperation analysis) are best contemplated once it has been established what work there is to allocate (all preceding dimensions). Likewise, decisions about what worker competencies are necessary (worker competencies analysis) are more sensibly made after it has been ascertained how work can be allocated among actors (social organization and cooperation analysis), for instance, between humans and machines. Thus, the earlier dimensions avoid commitments to decisions that are more naturally considered later during a work analysis.

What Is Unique about Cognitive Work Analysis?

A variety of techniques are available for analyzing work in complex sociotechnical systems. These techniques may be classified in terms of whether they offer normative, descriptive, or formative approaches to work analysis (Rasmussen, 1997; Vicente, 1999). To establish why cognitive work analysis is unique, I follow Rasmussen and Vicente in comparing this formative approach with normative and descriptive techniques.

Normative approaches *prescribe* how work *should be done* in a system. Task analysis techniques that specify sequences or timelines of tasks for workers to follow belong in this category (e.g., Kirwan and Ainsworth, 1992; Miller, 1953; Taylor, 1911). These techniques are well suited to highly stable systems, such as assembly-line operations, for which it is possible to anticipate the conditions that workers will confront. Ideal task sequences or timelines for handling these occurrences may be prespecified. By providing detailed guidance to workers about the ideal ways to perform tasks, normative approaches can optimize workload and minimize error.

Descriptive techniques focus on *describing* how work *is done* in a system. Cognitive task analysis techniques, which are intended for studying the cognitive nature of work, are of this type (e.g., Klein and Militello, 2001; Schraagen et al., 2000; Seamster, Redding, and Kaempf, 1997; Zsambok and Klein, 1997). They are most appropriate for investigating systems in which workers' responsibilities are characterized predominantly by cognitive rather than manual tasks. These workers have significant discretion for decision making or problem solving, and they usually do not—and cannot—rely on normative prescriptions of how work should be done. By studying the cognitive tasks of workers, and how those tasks are accomplished in real settings, descriptive techniques seek to establish how those tasks may be better supported.

Cognitive work analysis focuses on revealing how work *can be done* in a system. This framework recognizes that workers in complex sociotechnical systems have many possibilities for action, including what to do, when to do it, and how to do it. Furthermore, these workers have to contend with novel or unanticipated events, which pose immense threats to a system's effectiveness (Perrow, 1984; Rasmussen, 1969; Reason, 1990; Vicente, 1999). Therefore, instead of prescribing or describing how work should be done or is done in a system—in situations that are known or predictable—this framework focuses on identifying the constraints on actors. Within these constraints, workers can *form* a variety of work patterns, which is why cognitive work analysis is described as *form*ative. Workers can create innovative patterns of behavior, which is essential for dealing with unforeseen events, and they can adopt ways of working that suit their individual preferences. By focusing on constraints, cognitive work analysis leads to designs that support workers in adapting their behavior as they see fit while remaining within the boundaries of effective performance. In the following discussion, I lay out these arguments in more detail.

Cognitive work analysis adopts a formative orientation for four main reasons. First, the work demands of complex sociotechnical systems are primarily cognitive and social in nature, rather than physical or manual. Workers can exercise considerable judgment in decision making or problem solving as many possibilities for action are available to them. A worker may choose different behavioral trajectories to achieve the same goal or outcome because of variations in circumstances. Alternatively, due to intrinsic human variability, a worker may adopt different trajectories to obtain the same outcome, even when the circumstances remain unchanged. Furthermore, because of individual differences in humans, two workers may select different trajectories to achieve the same outcome in identical circumstances.

Second, it is not feasible at the time of design to prescribe or describe all of the possibilities for action available to workers. One reason for this is that these behavioral trajectories are too numerous for it to be practical to prescribe or describe the full set. Another is that workers may have to deal with events that are novel or cannot be predicted ahead of time. Examples of such events are an unforeseen change in the economy in the financial industry, an unpredicted rush order in the manufacturing industry, a new type of network failure in the communications industry, an unexpected allergic reaction of a patient to medication in the health care industry, and unanticipated multiple, interacting equipment failures in the nuclear power industry (Vicente, 1999).

As noted above, such events place significant pressure on a system's viability. Behavioral trajectories for handling these situations cannot be prescribed at the time of design because the goals to be achieved are unknown at that point. Furthermore, by definition, these trajectories cannot be described a priori, as work practices for dealing with well-known, familiar, or prior events, which can be identified during design, are unlikely to be suitable in the case of unexpected ones. Typically, workers must respond creatively to manage these situations successfully.

Third, to prescribe how work should be done in a system, analysts must have information about the properties of the devices (i.e., computer-based information systems such as interfaces or automation) used to perform that work. Consider, for instance, the fact that the behavioral trajectories necessary for preparing a meal with a microwave oven can only be specified if one has information about the features of that device. This type of information will not be available, though, if the reason for prescribing behavioral trajectories is to enable devices to be designed. In this case, analysts must make assumptions about the requisite functionality of new devices, rather than allowing the functionality to emerge from an analysis of work demands.

Fourth, descriptions of how work is done in a system are likely to reflect the properties of devices that workers already have. These descriptions may assimilate any unproductive practices stemming from those devices. At the same time, they may exclude potentially effective practices that are unsupported by those devices, as these practices cannot be observed. Another problem is that, because people are adaptive, a new device developed on the basis of how workers currently behave is likely to lead to changes in that behavior when it is introduced into a workplace. The ensuing behavior, however, may not be well supported by the new device. In these ways, a descriptive approach is limited by the functionality of existing devices.

Cognitive work analysis does not prescribe or describe how work should be done or is done in a system. As indicated above, normative or descriptive analyses are bound to be limited to a small subset of possibilities for action relative to the total number of possibilities. Consequently, they may result in designs that do not support workers in responding innovatively to variations in circumstances, including unpredictable events. These designs also may not support workers in following their individual preferences when this is appropriate. In addition to these limitations, normative and descriptive analyses are *device dependent*. This means that they may result in designs that are limited either by analysts' assumptions of the necessary functionality of new devices or by the functionality of existing devices.

Cognitive work analysis focuses on defining constraints. These constraints must be respected by actors regardless of the events they have to deal with, their individual preferences, or the details of proposed or existing devices. Within these constraints, actors still have many degrees of freedom for action. By basing designs on constraints, rather than prespecified or current ways of working, actors can be given the flexibility to adapt their behavior—in light of their judgment or preferences—without crossing the boundaries of effective performance. Interface designs, for example, may make the constraints on action visible to workers (Rasmussen and Vicente, 1989; Vicente, 2002; Vicente and Rasmussen, 1990, 1992) rather than just support them in following rules or procedures or in undertaking familiar work practices. Workers can then adopt new or different behavioral trajectories without violating those constraints. This approach, therefore, does not lead to designs inbuilt with fixed work practices but, instead, leads to designs that facilitate adaptation.

To avoid confusion, I note that although cognitive work analysis does not focus on producing descriptions of workers' behavior, it may involve studies of their behavior, as discussed in Chapter 8. The aim of such studies is to analyze the constraints on workers' behavior, not to develop descriptions of this behavior.

Designing for Adaptation

By designing for adaptation, cognitive work analysis promotes safety, productivity, and workers' health (Vicente, 1999). This approach to design is necessary for complex socio-technical systems because of the impact that novel events can have on their effectiveness (Perrow, 1984; Rasmussen, 1969; Reason, 1990; Vicente, 1999). As these events cannot be specified ahead of time, workers cannot be provided with preplanned solutions for handling them. Moreover, because workers have not experienced these events before, they do not have existing solutions for managing them. Instead, when such events occur, workers must use their expertise and ingenuity to improvise a solution *online* and *in real time* (Vicente, 1999) and thus *finish the design* (Rasmussen and Goodstein, 1987). By deliberately supporting workers in responding creatively to unanticipated events—without straying outside the boundaries of effective performance—cognitive work analysis provides an approach to design that fosters safety in organizations.

Designing for adaptation also enhances productivity. The rationale behind this claim is that complex sociotechnical systems are characterized by a high degree of computerization. This means that routine aspects of work for which algorithms can be written are usually automated. As a result, workers' responsibilities tend to involve intellectual activities, entailing a high degree of open-ended problem solving. Cognitive work analysis leads to greater productivity because designs are created with the express intention of facilitating workers in thinking innovatively in their jobs and supporting them in adopting new or different work practices as the situation demands. In addition, this framework is *device independent*. Designs based on constraints are not subject to analysts' assumptions about the required functionality of new devices or limited by the functionality of existing devices. Instead, such designs may incorporate novel functionality that emerges from the analysis of work demands.

Last, designing for adaptation promotes workers' health, including such factors as longevity and the absence of stress or disease. As discussed by Vicente (1999), several studies have shown that workers' health is influenced by the degree of control or decision latitude they have in their jobs. Specifically, workers with greater decision latitude tend to have better health. Workers with high decision latitude have the autonomy to decide how an activity should be performed. They have the freedom to follow their individual preferences when it is acceptable to do so and the discretion to cultivate their skills while doing their jobs. Cognitive work analysis purposefully provides workers with both opportunities by engendering designs that allow them many degrees of freedom for action.

Summary

Cognitive work analysis is a framework that defines the work demands of complex socio-technical systems in terms of the constraints on actors. It comprises five dimensions—work domain analysis, control task analysis, strategies analysis, social organization and cooperation analysis, and worker competencies analysis—for analyzing different types of constraints. By focusing on constraints, this framework seeks to establish how work *can be done* (formative approach) rather than how it *should be done* (normative approach) or *is done* currently (descriptive approach) in a system. Designs based on constraints promote flexibility or adaptation in the workplace. Workers can adopt new or different work practices to suit the demands of the situation or their individual preferences without violating the boundaries of effective performance. Such designs are also impervious to the properties of proposed or existing devices. By designing for adaptation, cognitive work analysis fosters safety, productivity, and workers' health.

This overview of cognitive work analysis provides ample context for the following chapters on work domain analysis. Readers are referred to the Appendix and the texts by Rasmussen (1986), Rasmussen et al. (1994), and Vicente (1999) for more comprehensive accounts of cognitive work analysis.

2

Work Domain Analysis

OVERVIEW   To appreciate the fundamentals of work domain analysis, it is necessary to comprehend the distinction between the analysis of actors' environments and that of their behavior. After presenting several analogies to demonstrate this distinction, this chapter explains why models of the environment—or *work domain*—are useful. The abstraction–decomposition space and the abstraction hierarchy, the main modeling tools for this form of analysis, are described. In addition, a work domain model of a home, a system that readers will recognize, is supplied as an example. This chapter then discusses two topics that are important for following the material in successive chapters. These are the classification of systems as causal or intentional and formats for work domain models.

What Is Work Domain Analysis?

Work domain analysis focuses on the constraints that are placed on actors by the functional structure of the field or environment in which they work. Terms like *field* or *environment* conjure up images of physical surroundings or conditions. Likewise, other terms used in discussing work domain analysis such as *landscape*, *map*, or *territory* are more naturally associated with the physical environment. However, workers usually operate within a social or cultural context as well. Work domain analysis, therefore, is concerned with the functional structure of the physical, social, or cultural environment of actors in a system, which places constraints on their behavior. Following Vicente (1999), I use *functional* to mean "action-relevant" (p. 155), or applicable for performing action, and *structure* to refer to "A relatively permanent relational property of a System" (p. 9).

   While work domain analysis is concerned with actors' environments, it is not concerned with their behavior. The distinction between actors' environments and their behavior is an important one. Whereas other techniques for work analysis focus on the latter, work domain analysis is concerned with the former. As this distinction can be difficult to appreciate in some contexts and to preserve in building work domain models, this section offers a number of analogies to illustrate it.

   Perhaps the simplest analogy for distinguishing between analyses of actors' environments and their behavior is Simon's (1981) parable about an ant on the beach (Rasmussen et al., 1994; Vicente, 1999). Figure 2.1 shows the path adopted by an ant in moving from one point to another on a surface that represents the beach. There are many paths the ant can take on the beach. Numerous paths are possible for reaching the same end point from the same starting point. In addition, many paths can be taken between a range of starting points and a particular end point as well as between a variety of different starting and end points. Work domain analysis is not concerned with the various paths the ant takes on the beach. Instead, its target is the functional structure of the beach, which places constraints on the ant's behavior or, in other words, shapes the paths the ant can take.

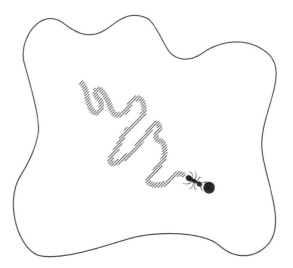

FIGURE 2.1
An illustration of the analogy of the ant on the beach. (From Vicente, K. J., 1999, *Cognitive work analysis: Toward safe, productive, and healthy computer-based work*, Erlbaum, Mahwah, NJ. With permission of Taylor and Francis Group. Permission conveyed through Copyright Clearance Center, Inc.)

Another analogy for distinguishing between analyses of actors' environments and their behavior is a comparison of street maps and directions (Vicente, 1999). Street maps depict the spatial layout of selected features in an environment, such as roads, bridges, traffic lights, roundabouts, bus stops, hospitals, fire stations, and shopping districts. Directions, on the other hand, specify the actions that are necessary for reaching a particular destination from a given starting point. Usually, many routes, and thus directions, are possible for traveling between any two locations. Consider, for instance, the number of routes that can be taken to get to destination point Z from starting point Y in Figure 2.2. Work domain analysis is not concerned with specifying actions for reaching any one point from another. Rather, it is aimed at mapping the functional structure of the landscape, which reveals the constraints on action, as well as the possibilities for action, for traversing between any two points.

The distinction between analyzing actors' environments and their behavior may be illustrated more directly with the analysis of a home. Consider the number of paths or trajectories that are possible for entering a home (Figure 2.3). An inhabitant may enter a home through its front door, through the back door or, in the event of forgetting the house keys, by breaking in through any one of its windows. Work domain analysis does not involve prescribing or describing the trajectories that inhabitants adopt. Its concern is with developing a representation of the functional structure of the home, which reveals the constraints on the trajectories that are possible.

Like the preceding analogies, the home example emphasizes the constraints associated with the physical environment. However, the social or cultural context can place constraints on actors as well. For instance, in a home, constraints relating to time or money may shape actors' behavior. If an inhabitant forgets the house keys, he or she may decide to save time by breaking in through a window or save money by waiting outside until other inhabitants arrive home. Work domain analysis is not concerned with specifying such trajectories for entering a home, but it is concerned with modeling the functional structure of the physical, social, or cultural context of a home, which shapes the trajectories that inhabitants can adopt.

FIGURE 2.2
Many routes, and thus actions, are possible for reaching destination point Z from starting point Y, especially considering alternative means of travel (e.g., walking, cycling, swimming, driving a car, catching a bus).

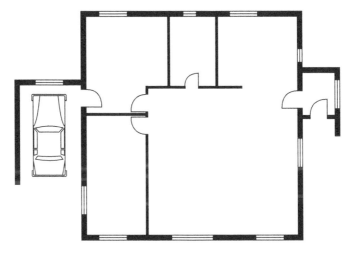

FIGURE 2.3
Several trajectories are possible for entering a home.

Finally, the distinction between analyses of actors' environments and their behavior may be demonstrated precisely with a study of professional electronic technicians in the industry of computer repair (Rasmussen and Jensen, 1974). Figure 2.4 depicts a technician's problem-solving trajectory in diagnosing faults in computer equipment; each numbered node represents a verbal statement made by the technician in reasoning about the task. This trajectory is superimposed onto an abstraction–decomposition space, which represents the 'field' on which the technician performed the fault diagnosis task, that is, the computer equipment. Thus, the trajectory reflects those aspects of the computer equipment that the technician was reasoning about in performing fault diagnosis, such as its purposes, functions, and components.

Using the analogy of the ant on the beach, the technician's problem-solving trajectory may be likened to the path of the ant, whereas the underlying abstraction–decomposition space may be likened to the beach. Moreover, Rasmussen and Jensen (1974) found that, like the ant on the beach, technicians could adopt many different trajectories, even when they were performing the same task of diagnosing faults in computer equipment. As before, however, work domain analysis does not focus on specifying these trajectories. Instead, its objective is to model the functional structure of the computer equipment, which places

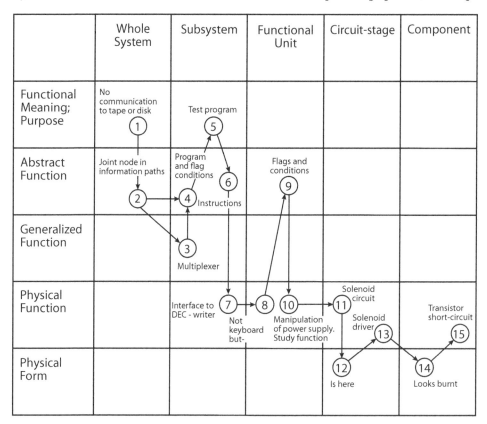

FIGURE 2.4
An electronic technician's problem-solving trajectory superimposed onto an abstraction–decomposition space. (Adapted from Rasmussen, J., 1988, A cognitive engineering approach to the modeling of decision making and its organization in: Process control, emergency management, CAD/CAM, office systems, and library systems, In Rouse, W. B., Ed., 1988, *Advances in man-machine systems research*, 4, JAI Press, Greenwich, CT, 165–243. Copyright © 1988 JAI Press. With permission from Elsevier.)

constraints on the trajectories that are possible. As highlighted by Figure 2.4, the abstraction–decomposition space provides a tool for developing such a model.

Why Are Models of the Work Domain Useful?

The preceding analogies show that work domain analysis is concerned with the functional structure of actors' environments. Given that other work analysis techniques focus on actors' behavior, it is worth laying out explicitly why models of the work domain are useful. To do this, I draw on the analogies presented previously.

First, as the functional structure of the environment places constraints on actors, a work domain model is useful because it identifies exactly what these constraints are. A landscape map depicts the constraints on the routes for traveling between any two locations. Alternatively, a representation of the physical, social, and cultural context of a home shows the constraints on the trajectories for entering the home.

Second, although the functional structure of the environment imposes constraints on actors, it still affords them many possibilities for action. A work domain model is beneficial, therefore, because it reveals not only constraints but also possibilities for action. A landscape map shows all of the available routes for reaching any one location from another, and a model of a home's environment portrays all of the viable trajectories for entering the home. It is worth emphasizing that a work domain model only shows actors what is possible. Actors have the flexibility to choose which actions to adopt on any occasion.

Third, because the functional structure of the environment places constraints on actors, it is not possible to understand their behavior without an awareness of their environment. As observed by Simon (1981), the path of an ant on the beach, which is irregular, complex, and difficult to describe, cannot be understood solely by studying the properties of the ant. A representation of the beach is also necessary because the ant's path is shaped by its properties as well. Likewise, without appreciating the physical, social, and cultural context of a home, it is not possible to understand why an inhabitant might choose to gain entry into the house by breaking a window on one occasion and by waiting until other inhabitants arrive home on another. Thus, an additional advantage of a work domain model is that, by defining the constraints on actors, it offers a way of understanding the rationale for their behavior.

Fourth, the functional structure of the environment remains relatively constant, unlike actors' behavior. Actors may take many different routes for traveling between any two locations, but the landscape stays the same. Similarly, electronic technicians may adopt a variety of problem-solving trajectories for diagnosing faults in computer equipment, but the functional structure of this apparatus does not change. Thus, while actors' behavior varies in accordance with the situation, their environment is comparatively invariant. A work domain model is useful, therefore, because it is *event independent*. This means that it identifies the constraints on actors' behavior, or their possibilities for action, regardless of the situation.

Last, because the functional structure of the environment is event independent, it provides a basis for reasoning about any situation, including those that are novel or unanticipated. As Vicente (1999) points out, people can use a map to derive a route from any starting point (e.g., home, office) to a variety of destinations (e.g., shops, airport). Even in the case of an unanticipated event, such as a traffic jam, a map can be used to formulate an alternative route to the desired destination. A map also allows people to recover from errors. If a person takes a wrong turn, for instance, he or she can get back on track to the desired destination. Moreover, a map allows people to select routes that suit their individual preferences; one might choose the most efficient route between two locations in one instance and the most scenic route between the same two locations another time. Hence, a work domain model is beneficial because it leads to designs that support actors in dealing with a range of situations, including unforeseen events.

The Abstraction–Decomposition Space

The abstraction–decomposition space is a tool for developing a model of the functional structure of the physical, social, or cultural environment of actors in a system, which places constraints on their behavior. This tool, which takes the form of a matrix, represents a work domain along two orthogonal dimensions (Figure 2.5). The abstraction dimension, along the vertical axis, comprises several qualitatively distinct concepts for modeling the structural properties of the environment. These levels are linked by structural means–ends relations. Along the horizontal axis, the decomposition dimension constitutes a set of levels of detail or resolution for modeling the structural properties of the environment. The connections between these levels are part–whole relations. Within the cells of the matrix, the structural properties of the environment are represented at different levels of abstraction and decomposition. These properties place constraints on actors.

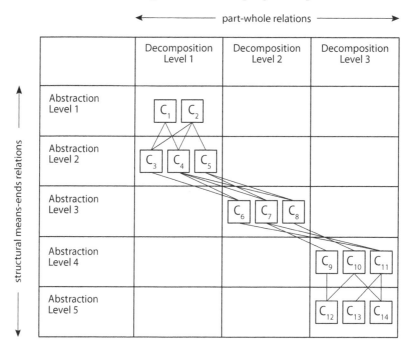

FIGURE 2.5

The abstraction–decomposition space. The nodes within the cells of the matrix signify constraints. C, constraint.

The various elements of the abstraction–decomposition space are explained in depth later in this chapter.

Although the abstraction–decomposition space in Figure 2.5 has five levels of abstraction and three levels of decomposition, the number of levels may vary. The types and labels of the levels may also vary, which is why generic levels are used in Figure 2.5. In Chapters 3 and 4, I address matters regarding the number, types, and labels of the levels of abstraction and decomposition to include in a work domain model.

Instead of modeling a work domain with an abstraction–decomposition space, analysts sometimes use an abstraction hierarchy. Figure 2.6 presents the constraints from the representation in Figure 2.5 in the form of an abstraction hierarchy. Unlike the abstraction–decomposition space, the abstraction hierarchy does not indicate the levels of decomposition to which the constraints in a model belong. A more detailed discussion of formats for presenting work domain models is provided at the end of this chapter.

Abstraction–Decomposition Space of a Home

This section presents an abstraction–decomposition space of a home. This model is used to exemplify the concepts and guidelines for work domain analysis as well as its modeling tools. Instead of a home, many other systems could have been chosen for illustration, but a home was selected because readers will find it familiar.

Figure 2.7 displays the abstraction–decomposition space of a home. This model is in a tabular format, whereas the abstraction–decomposition space in Figure 2.5 demonstrates a graphical one. The relationships between these and other formats for work domain models are explained later in this chapter.

The model of a home has five levels of abstraction: functional purposes, value and priority measures, purpose-related functions, object-related processes, and physical objects. As discussed in Chapter 3, these levels are the same as those described by Rasmussen and his colleagues (Rasmussen, 1979, 1985, 1986; Rasmussen et al., 1994). The labels for these levels, which were also formulated by Rasmussen, are documented by Reising (2000). These labels may be better suited to *intentional systems*, in which actors' behavior is constrained primarily by social laws, conventions, or values, rather than *causal systems*, in which actors'

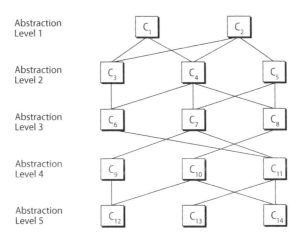

FIGURE 2.6
The abstraction hierarchy. C, constraint.

	Whole House	Rooms or Subspaces	Contents
Functional Purposes	Well-being, Environmental protection, Residential laws and regulations etc.	*Functional purposes of rooms or subspaces*	*Functional purposes of contents*
Value and Priority Measures	Total income – total expenses = savings, Time, Health, Hygiene, Pleasure, Conservation of natural resources etc.	*Value and priority measures of rooms or subspaces*	*Value and priority measures of contents*
Purpose-related Functions	*Purpose-related functions of whole house as a single entity*	Preparation of meals and beverages, Dining, Recreation, Rest, Exercise, Personal care and grooming, Housework, Maintenance, Professional work etc.	*Purpose-related functions of contents*
Object-related Processes	*Functional capabilities or limitations of whole house as a single entity*	*Functional capabilities or limitations of rooms or subspaces*	Washing capacity, Cooling capacity, Heating capacity, Food and beverage handling capacity, Serving and seating capacity, Reception and audiovisual capacity, Audio playback capacity, Seating capacity, Serving and exhibiting capacity, Sleeping capacity, Showering capacity, Bathing capacity, Brushing capacity, Drying capacity, Transmission and audiovisual capacity, Data handling capacity etc.
Physical Objects	*Physical form of whole house as a single entity*	*Physical form of rooms or subspaces*	Dishwasher, Refrigerator, Stove, Utensils, Table, Television, Compact disc player, Armchair, Coffee table, Bed, Shower, Bath, Toothbrush, Towel, Telephone, Computer etc.

FIGURE 2.7
Abstraction–decomposition space of a home.

behavior is constrained primarily by physical or natural laws. Following Rasmussen et al., I also describe the concepts underlying these levels in terms that may be better suited to intentional systems. This rendering of the abstraction dimension, therefore, complements that in the texts by Rasmussen (1986) and Vicente (1999), which is arguably better suited to causal systems. The distinction between intentional and causal systems is dealt with at length toward the end of this chapter.

The home model has three levels of decomposition: whole house, rooms or subspaces, and contents. Unlike the abstraction dimension, the decomposition dimension has no standard set of levels, although the general principle is that this dimension represents a system at finer levels of detail when moving from left to right.

The shaded cells along the diagonal of the model contain examples of constraints. These cells offer the most useful information for illustrating the concepts and guidelines for work domain analysis. The remaining cells contain an overview of their potential content.

The home model is discussed comprehensively during the course of this book. When constraints from the model are mentioned in the text, they are shown in *italics*. Readers may wish to make a copy of the model, which they can refer to easily as they work through the text.

The rest of this section documents a number of important points about the scope of the abstraction–decomposition space of a home. The first observation is that the model is of a particular home, not every home. Accordingly, when I refer to the model of *a* home, readers need to bear in mind that the allusion is to a specific home rather than a generic one. Work domain analysis is typically performed on individual systems, not general classes of systems.

The model in Figure 2.7 is not the only possible representation of the home. There are many ways to model the same system, depending on the purpose of the analysis, as is explained in Chapter 8. The purpose of this home model is to illustrate the material on work domain analysis with examples from an intentional system. Hence, the boundaries of the analysis focus on the home as a social system rather than a physical one. The use of the term *home* rather than *house* to refer to the model reflects this focus. By selecting an intentional system for illustration, the examples provided here complement Vicente's (1999) case study of DURESS II, a thermalhydraulic microworld simulation, which is a causal system.

Another important observation is that examining the home model for usefulness or comprehensiveness with respect to any application other than that for which it was developed is unproductive. This model is sufficient for explaining work domain analysis to readers. If a model of a home had been developed with a specific design goal in mind, for instance, the resulting content would have been narrower and more detailed. Such a model would not have been broad enough to illustrate a range of discussions about work domain analysis and would have been unnecessarily complex for explaining this technique to readers. Chapters 9, 10, and 11 provide comprehensive case studies of work domain models produced for particular applications.

Readers will also find it difficult to evaluate the abstraction–decomposition space of a home for correctness. One reason for this is that the labels of the constraints in this model are not always self-explanatory. For example, the label of one of the functional purposes is *well-being*. This label alone does not convey the intended meaning of this term within the model, which is a mental and physical state characterized by health, happiness, and prosperity. As discussed in Chapter 7, a glossary that describes the constraints in a work domain model in greater detail than its graphical or tabular format is often useful. Fuller descriptions of the constraints in the model of a home are provided in the text where necessary.

Not all of the information relevant to a design problem is necessarily captured by work domain analysis. If readers find themselves questioning whether the home model is missing important information, they need to bear in mind that it may be the focus of one of the other dimensions of cognitive work analysis. The key concepts of the remaining dimensions are dealt with in the Appendix.

As a home is a system that is well known to readers, the examples from this model may at times seem simplistic. A home was chosen deliberately as the main system for illustration so that readers would not experience needless difficulty in following the material on work domain analysis, as may have been the case with a more specialized system. This material is often also exemplified with models of a variety of other systems, including libraries, hospitals, petrochemical plants, and nuclear power plants. Moreover, Chapters 9, 10, and 11, which describe the application of work domain analysis to military systems, provide more complex illustrations of this technique.

In what follows, I use the model of a home to explain more thoroughly the key features of the abstraction–decomposition space, which subsumes the abstraction hierarchy. These features are the abstraction and decomposition dimensions, structural means–ends and part–whole relations, and constraints. Why the abstraction–decomposition space can be used for modeling work domains and the relationship of this type of model to human reasoning are addressed as well.

Abstraction Dimension

The abstraction dimension of a work domain model is defined by a set of qualitatively different concepts for modeling the functional structure of the environment. Typically, these concepts span from the purposes of a system to its physical form.

The model of a home (Figure 2.7) has five levels of abstraction, which are suitable for modeling a variety of systems. The functional purposes level represents the objectives of a system (e.g., *well-being*) and the external constraints on its operation (e.g., *residential laws and regulations*). At the value and priority measures level, the criteria that must be respected (e.g., *total income – total expenses = savings*) for a system to achieve its functional purposes are shown. The next level, purpose-related functions, portrays the functions a system must be capable of supporting (e.g., *preparation of meals and beverages*) so that it achieves its functional purposes. The object-related processes level specifies the functional processes or functional capabilities or limitations (e.g., *washing capacity*, including load limits, washing cycles, and temperature range) of physical objects in a system. The last level, physical objects, presents the physical objects themselves (e.g., *dishwasher*). This set of five levels of abstraction is described comprehensively in Chapter 3.

Generally, higher levels of abstraction in a work domain model represent *purposive* concepts. These are concepts that relate to the reasons for a system's existence or design. In contrast, lower levels of abstraction model *physical* concepts, which relate to a system's physical resources.

Although the different levels of abstraction in a work domain model portray the same system, the concepts at each level are qualitatively distinct. Therefore, examining a system at different levels is like viewing it through different conceptual lenses. In the model of a home, the five levels of abstraction provide a view of this system's functional purposes (e.g., *well-being, environmental protection*); value and priority measures (e.g., *pleasure, time*); purpose-related functions (e.g., *recreation, professional work*); object-related processes (e.g., *reception and audiovisual capacity, seating capacity*); and physical objects (e.g., *television, armchair*). Each level is associated with a unique "language" for depicting the view from that level (Vicente, 1999, p. 171). When moving up and down the abstraction dimension, then, the conceptual lens at each level *changes* the view of a system rather than just adds or removes details from the same view.

Decomposition Dimension and Part–Whole Relations

The decomposition dimension of a work domain model comprises a number of levels of detail or resolution for modeling the functional structure of the environment. The model of a home (Figure 2.7) has three levels of decomposition, specifically whole house, rooms or subspaces, and contents. At the first level, the home is modeled at the granularity of the entire house as a single entity. The next two levels model the home at the granularity of its rooms or subspaces and contents, respectively.

As a rule, higher levels of decomposition in a work domain model provide coarse representations of a system. Lower levels, in contrast, provide fine-grained ones. In the model of a home, the whole house level presents a coarse representation of this system, whereas the contents level presents a fine-grained depiction.

The levels of decomposition in a work domain model are connected by *part–whole relations*. This means that entities at lower levels are functional parts of those at higher levels. Conversely, entities at higher levels are functional wholes of those at lower levels. In the home model, contents are functional parts of rooms or subspaces, and rooms or subspaces are functional parts of the whole house. Alternatively, the whole house is composed of rooms or subspaces, and rooms or subspaces are composed of contents. When moving across this dimension from right to left, then, the parts of a system are aggregated into functional wholes, whereas when moving across this dimension from left to right, the whole system is decomposed into its functional parts.

While the levels of decomposition in a work domain model portray the same system, each level provides a different resolution for viewing that system. In the model of a home, the three levels of decomposition provide a view of this system at the resolutions of the entire house, its rooms or subspaces, and its contents. Therefore, as observed by Vicente (1999), moving across this dimension from left to right is like "zooming in" because each consecutive level offers a more detailed view of the same system (p. 158). In contrast, moving from right to left is like "zooming out" because each consecutive level offers a less detailed view of the same system. Accordingly, whereas the abstraction dimension of a work domain model represents the conceptual lenses for viewing a system, the decomposition dimension represents the resolutions of those conceptual lenses.

Constraints

The cells of the abstraction–decomposition space are populated with representations of the functional structure of the environment, which places constraints on actors. Each representation is at a specific level of abstraction and decomposition. In the home model (Figure 2.7), the constraints on actors are depicted in the following terms: functional purposes and value and priority measures at the whole house level of decomposition, purpose-related functions at the rooms or subspaces level of decomposition, and object-related processes and physical objects at the contents level of decomposition.

When the cells of an abstraction–decomposition space are populated fully, each cell offers a complete but different representation of the same system (Vicente, 1999). Consider the top left and bottom right cells in the model of a home. In its entirety, either cell would provide a complete representation of the same home. The top left cell would describe all of the functional purposes of this system at the whole house level of decomposition, and the bottom right cell would describe all of its physical objects at the contents level of decomposition. However, the two cells would present different depictions of the same home. The top left cell would provide a purposive model of the entire home (i.e., a coarse, purposive model), whereas the bottom right cell would provide a physical model of its individual contents (i.e., a fine-grained, physical model).

Alternatively, consider the bottom left and right cells in the home model. As before, in its entirety, either cell would contain a complete representation of the same system. The bottom left cell would depict all of the physical objects of the home at the whole house level of decomposition (i.e., the entire house as a single entity), and the bottom right cell would depict all of its physical objects at the contents level of decomposition. Even though both cells would provide a physical model of the same home, the two representations would be different. The bottom left cell would offer a physical model of the entire home (i.e., a coarse, physical model), whereas the bottom right cell would offer a physical model of its individual contents (i.e., a fine-grained, physical model).

While it is possible to populate every cell of an abstraction–decomposition space with constraints, it is rarely productive or efficient to do so (Miller and Vicente, 1998). Instead, only those cells that offer the most meaningful or useful information about a system, given the purpose of the analysis, are worthy of detailed attention. Typically, cells that fall around the diagonal of a model from top left to bottom right provide the most useful representations of a system. This is because coarse levels of resolution are useful for considering a system's overall purposes, whereas fine levels of resolution are best for reasoning about its individual physical objects. In the model of a home, only five cells around the diagonal are populated with constraints.

Structural Means–Ends Relations

The abstraction dimension of a work domain model is defined by *structural means–ends relations*. Concepts at lower levels represent structural *means* for achieving the ends at higher levels. Conversely, concepts at higher levels represent structural *ends* that can be achieved by the means at lower levels.

Structural means–ends relations are an essential characteristic of work domain models, and they are unique to this form of representation. As inexperienced analysts often have difficulty distinguishing such links from other types of hierarchical relations, it is worth listing several examples of this critical feature, from every level of abstraction, using the model of a home (Figure 2.7). Moving from bottom to top, a *dishwasher* is a means by which *washing capacity* can be achieved; *washing capacity* is an end that can be achieved with a *dishwasher*. In the same way, *washing capacity* provides a means for accomplishing the end of *housework*, and *housework* provides a means for attaining the end of *hygiene*. Last, *hygiene* is a means by which *well-being* can be realized; *well-being* is an end that can be realized with *hygiene*.

The structural means–ends relations in a work domain model may be portrayed as a how-what-why triad, which reflects how workers can formulate their work demands in relation to the context in which they are situated. The level of abstraction at which workers define their actions at any time signifies *what* needs to be done. In Figure 2.8, this is represented by Purpose-related Function A. At the level below this, workers can examine their actions in terms of the means for achievement or *how* Purpose-related Function A can be done. Object-related Processes B and C represent the means by which Purpose-related Function A can be accomplished. At the level above, workers can examine their actions in

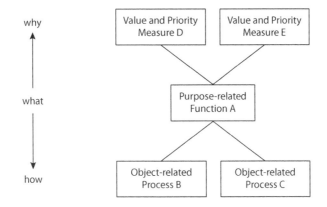

FIGURE 2.8
How-what-why triad of structural means–ends relations.

terms of the ends to be achieved or *why* Purpose-related Function A should be done. Value and Priority Measures D and E represent the ends that can be attained with Purpose-related Function A. As shown in Figure 2.9, the how-what-why triad can be applied by starting at other levels of abstraction.

It is important to distinguish structural means–ends relations from action means–ends relations, which do not belong in a work domain model. Structural means–ends relations describe the properties of the environment that are necessary for achieving an end. In contrast, action means–ends relations describe the tasks or activities that are relevant for achieving an end. To recast an example provided by Vicente (1999), a *furnace* and a *fireplace* are structural means for attaining *warmth* because they are properties of the environment (Figure 2.10a). However, *going down to the basement* and *lighting a fire in the fireplace* are action means for attaining *warmth* because they are tasks or activities (Figure 2.10b). Given that the abstraction dimension comprises concepts pertaining to the functional structure of actors' environments, and not their behavior, it is characterized by structural means–ends relations, not action means–ends relations. It is also worth emphasizing that the functional

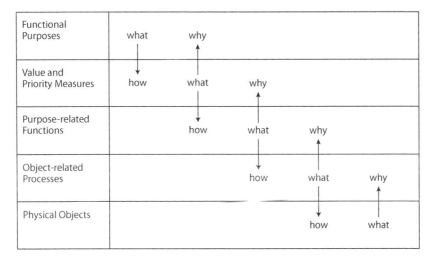

FIGURE 2.9
The how-what-why triad applied to multiple levels of abstraction.

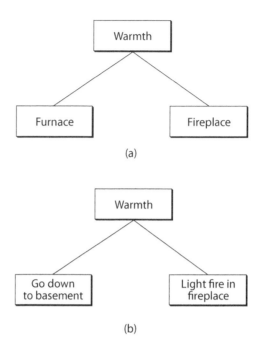

FIGURE 2.10
Illustration of the distinction between (a) structural means–ends relations and (b) action means–ends relations.

structure of the environment places constraints on the tasks or activities that are necessary for achieving an end. For instance, different tasks or activities are needed depending on whether one uses a furnace or a fireplace to achieve warmth.

Complex sociotechnical systems are characterized by many-to-many structural means–ends relations. A means at a lower level of abstraction can often be used to accomplish several ends at a higher level (Figure 2.11a). In the model of a home (Figure 2.7), *personal care and grooming* is a means for realizing *hygiene, conservation of natural resources, pleasure*, and *total income – total expenses = savings*. (To explain some of the relationships that may not be obvious from the labels, *conservation of natural resources* can be achieved by minimizing the amount of water one uses for *personal care and grooming*, for example, in the shower. Limiting water usage would also lower utility bills, thereby respecting the relationship *total income – total expenses = savings*.) Conversely, an end at a higher level of abstraction can often be accomplished by several means at a lower level (Figure 2.11b). For instance, *personal care and grooming* can be achieved with the *showering capacity* of a *shower*, the *bathing capacity* of a *bath*, the *brushing capacity* of a *toothbrush*, or the *drying capacity* of a *towel*.

Many-to-many structural means–ends relations reflect the numerous possibilities for action available to actors and, accordingly, the complexity of their decisions or choices (Rasmussen et al., 1994). When multiple ends can be achieved by the same means, actors need to consider how using a means to fulfill one end will affect the others. In the context of the home model, actors must take into account how the use of *personal care and grooming* to attain *hygiene* and *pleasure* will affect the criteria of *conservation of natural resources* and *total income – total expenses = savings*. In addition, if an end can be achieved by several means, actors must judge which means is best for achieving the end in question. With respect to the home model, actors may need to consider whether *personal care*

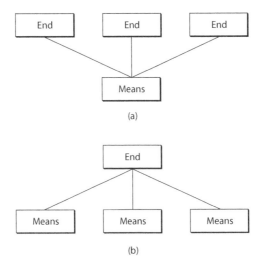

(a)

(b)

FIGURE 2.11
Many-to-many structural means–ends relations: (a) A means at a lower level can be used to attain several ends at a higher level; (b) an end at a higher level can be satisfied by several means at a lower level.

and grooming is best attained with the *showering capacity* of a *shower* or the *bathing capacity* of a *bath*.

Actors must take into account both the desired and undesired effects of their decisions or choices. In relation to the home model, employing a *bath* and its *bathing capacity* for *personal care and grooming* may achieve both *hygiene* and *pleasure*, but it may have undesired effects on the criteria of *conservation of natural resources* or *total income – total expenses = savings*. Alternatively, employing a *shower* and its *showering capacity* for *personal care and grooming* may have desired effects on *hygiene, conservation of natural resources*, and *total income – total expenses = savings*. It may not, however, have the required effects on *pleasure*.

The set of structural means–ends relations that must be considered during routine or familiar situations is usually well established (Rasmussen et al., 1994). For example, the relationships that inhabitants must be mindful of in deciding whether to have a bath or a shower or in deciding how much time to spend in the shower under typical conditions will be well known to them. Decision making in these situations is relatively straightforward. In contrast, during novel or unanticipated situations, all possible structural means–ends relations must be evaluated explicitly in light of the current circumstances (Rasmussen et al., 1994). For instance, in the event of a sharp increase in water prices, inhabitants may need to assess all potential relationships to determine how they can respect the criterion of *total income – total expenses = savings* while still satisfying other ends such as *hygiene* and *pleasure*. Previously unrecognized structural means–ends relations may be revealed in these situations, and decision making can be challenging.

A Tool for Modeling Work Domains

The abstraction–decomposition space provides a tool for modeling work domains because it represents the functional structure of the environment of actors, not their behavior. As this is a concept that is often misunderstood, it is worth making this point explicitly for each level of abstraction. In the model of a home (Figure 2.7), the functional purposes level describes the objectives of this system (e.g., *well-being*) and the external constraints on its operation (e.g., *residential laws and regulations*), not actors' behavior. Likewise, the next four levels of abstraction do not represent actors' behavior. The second level, value and priority measures, identifies the criteria that must be respected (e.g., *pleasure*) for the home to achieve its functional purposes, and the third level, purpose-related functions, describes the functions this system must be capable of supporting (e.g., *recreation*) so that it fulfills its functional purposes. The object-related processes level represents the functional processes or functional capabilities or limitations (e.g., *reception and audiovisual capacity*) of physical objects of the home, and the lowest level portrays those physical objects (e.g., *television*). The abstraction–decomposition space, therefore, models the environment of actors in terms of the functional purposes, value and priority measures, purpose-related functions, object-related processes, and physical objects of a system, as this is the context within which they perform their work.

By modeling the functional structure of actors' environments, the abstraction–decomposition space identifies the constraints on their behavior. In the home model, the functional purpose of *well-being*, the value and priority measures of *total income – total expenses = savings*, *health*, or *pleasure*, and the purpose-related function of *preparation of meals and beverages* place constraints on the types of meals that inhabitants make. In the same way, at the next two levels of abstraction, a *stove* and its *heating capacity*, *utensils* and their *food and beverage handling capacity*, and a *refrigerator* and its *cooling capacity* impose constraints on inhabitants' meal choices. In sum, the abstraction–decomposition space identifies the constraints on actors' behavior in terms of the functional purposes, value and priority measures, and purpose-related functions they must fulfill with a given set of physical resources.

Similarly, by providing a model of the environment of actors, the abstraction–decomposition space reveals the multiple possibilities for action open to them. In relation to the home model, inhabitants have many options for preparing meals that satisfy the purpose-related function of *preparation of meals and beverages*, the value and priority measures of *total income – total expenses = savings*, *health*, or *pleasure*, and the functional purpose of *well-being*. Likewise, many options are available to inhabitants for preparing meals with physical resources such as a *stove* and its *heating capacity*, *utensils* and their *food and beverage handling capacity*, and a *refrigerator* and its *cooling capacity*. In general, then, actors have numerous possibilities for action to satisfy a system's functional purposes, value and priority measures, and purpose-related functions with the existing physical resources. This means that actors usually have the flexibility to choose which actions to adopt on any occasion.

The abstraction–decomposition space also offers a way of understanding the rationale for actors' behavior. In the model of a home, value and priority measures such as *total income – total expenses = savings* and *time* may explain why inhabitants sometimes prepare meals at home and sometimes eat meals at a restaurant. In the same way, the size of a *dishwasher*, which affects its *washing capacity*, may explain why actors wash some items by hand. Accordingly, the rationale for actors' behavior may be understood not only in terms of the functional purposes, value and priority measures, and purpose-related functions they must realize, but also in terms of the physical objects that are available to them and the functional capabilities or limitations of those objects.

Finally, because the abstraction–decomposition space models the functional structure of the environment, it is relevant across a range of situations. A home's functional purposes (e.g., *well-being*); value and priority measures (e.g., *hygiene*); purpose-related functions (e.g., *personal care and grooming*); object-related processes (e.g., *showering capacity*, *data handling capacity*); and physical objects (e.g., *shower*, *computer*) are relatively constant, unlike actors' behavior. The abstraction–decomposition space, therefore, is event independent—that is, it identifies the constraints on actors, as well as their possibilities for action, irrespective of the situation.

Given that the abstraction–decomposition space is event independent, it provides actors with a basis for reasoning about any situation, including unforeseen events. Every situation can be considered in terms of what actions are required to fulfill the functional purposes, value and priority measures, and purpose-related functions of the relevant system with the object-related processes and physical objects that are available. In the context of the home model, a value and priority measure such as *total income – total expenses = savings* provides a principle for meal planning, whether these are routine meals or a rare dinner party. Inhabitants can plan meals that respect this principle while still satisfying their individual preferences. In addition, this principle can be used for reasoning about the possibilities for handling unanticipated situations, such as having unexpected guests at a dinner party, or dealing with errors, such as forgetting to cater for vegetarian guests. For example, actors can assess whether they can afford to order more food from a restaurant or whether other strategies are necessary. The abstraction–decomposition space, therefore, can support actors in managing a range of situations, including novel events.

Relevance to Human Reasoning

Rasmussen (1979, 1985) formulated the abstraction–decomposition space on the basis of a variety of studies of human reasoning. These included studies performed by other researchers (e.g., de Groot, 1965; Duncker, 1945; Frijda and de Groot, 1981) as well as his own (Rasmussen, 1974, 1976, 1979, 1985, 1986; Rasmussen and Jensen, 1974; Vicente and Rasmussen, 1992). Unlike laboratory investigations of human reasoning, which involved naïve subjects carrying out well-defined but unfamiliar tasks (e.g., Newell and Simon, 1972), Rasmussen and his colleagues studied experienced workers in their actual work settings doing their normal jobs.

In one of these field studies, Rasmussen and Jensen (1974) investigated the way in which professional technicians troubleshoot faults in electronic equipment. Some of the aims of this research were to examine the problem-solving strategies of workers and to investigate the structural properties of the system that was the object of their activities (i.e., the electronic equipment). These observations would contribute to establishing how the system can be represented in a model that is both compatible with human problem solving and event independent. With such a work domain model, workers should be able to reason about the system in a wide range of situations, including unforeseen events, and thus deal with these situations more effectively.

The field study included eight different types of instruments, each with a particular fault, and six professional technicians. The investigation was based on verbal protocol methodology, so technicians were required to verbalize their problem-solving processes as they set about their troubleshooting tasks. A total of 45 cases were recorded and transcribed, although only 30 were subjected to detailed analysis.

The data analysis involved the development of a preliminary coding scheme, which was used to analyze the verbal protocols. The protocols were then reviewed to determine if the information in them was well captured by the coding scheme. Any discrepancies led to changes in the coding scheme until it stabilized. The verbal protocols were subsequently analyzed with the final coding scheme.

On the basis of the results of this and other studies, a number of patterns were identified in the way workers reason about complex systems during problem solving in a range of situations. First, workers reason at different levels of *abstraction* while performing their jobs (Figure 2.12). Specifically, they spontaneously shift their view of a system from purposive concepts to physical concepts in order to match their task demands. They adopt purposive concepts when considering their task demands in terms of the purposes to be achieved and physical concepts when thinking about their task demands in relation to the resources that are available.

Workers also reason at different levels of *decomposition* while performing their jobs. That is, they spontaneously shift their view of a system from coarse levels to fine-grained levels of resolution in order to match their task demands. They adopt coarse models when considering their task demands at the level of the entire system and fine-grained models when reflecting on their task demands at the level of the system's components.

Furthermore, when the conceptual lens through which workers view a system changes from purposive to physical, the level of resolution at which they view the system also changes from coarse to fine grained. As a result, workers tend to adopt models along the diagonal of this reasoning space, which indicates that these representations generally offer the most meaningful or useful views of a system (see, for example, the shaded cells in Figure 2.12). Hence, although the abstraction and decomposition dimensions are conceptually orthogonal, they are actually coupled in practice (Vicente, 1999).

Last, this reasoning space is event independent. This means that, although workers may adopt diverse problem-solving paths in the same situation or unique problem-solving paths across a range of situations, all of these trajectories can be mapped onto, or can be explained by, this reasoning space. (Figure 2.4 shows how an electronic technician's problem-solving trajectory while diagnosing faults in computer equipment can be mapped onto this type of representation.)

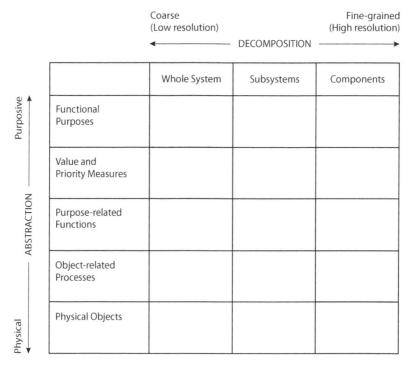

FIGURE 2.12
The abstraction–decomposition space is consistent with the characteristics of human problem solving.

On the basis of these findings, Rasmussen (1979, 1985) proposed the abstraction–decomposition space as a framework for representing complex systems in a way that is both consistent with the characteristics of human problem solving and event independent. Such a work domain model can support human reasoning about complex systems in a range of situations, including those that are novel or unforeseen.

Since Rasmussen (1979, 1985) put forward the abstraction–decomposition space, other researchers have shown that actors' problem-solving trajectories can be mapped onto models developed with this tool. One study by Itoh, Yoshimura, Ohtsuka, and Masuda (1990) was conducted with a team of nuclear power plant operators in a high-fidelity simulator. The operators were required to perform both supervisory and control tasks in an unfamiliar scenario and a familiar but complex scenario. Another study by Vicente, Christoffersen, and Pereklita (1995) involved subjects with a broad range of expertise who were required to participate in a thermalhydraulic microworld simulation and diagnose both normal and abnormal system behavior in five scenarios involving unfamiliar events.

In both of these studies, participants' verbal protocols during the simulations were mapped onto an abstraction–decomposition space of the relevant system. Although participants adopted a variety of problem-solving paths, even when they were performing the same task, all of these trajectories were accommodated by the relevant model. Itoh et al. (1990) found that the trajectories tended to fall around the diagonal of the model in their study. Vicente et al. (1995) demonstrated that the trajectories that fell around the diagonal of their model were strongly correlated with effective performance. These trajectories were also positively correlated with participants' prior knowledge or experience of the thermalhydraulic microworld simulation and the interface used in the experiment. This result suggests that experienced actors are more likely than inexperienced actors to adopt trajectories that fall around the diagonal of the abstraction–decomposition space.

Such studies demonstrate that the abstraction–decomposition space provides a tool for modeling work domains in a way that is consistent with actors' problem-solving or reasoning processes. In this sense, it may be viewed as representing the fundamental *problem space* or *reasoning space* of actors. Reasoning at different levels of abstraction and decomposition allows actors to cope with complexity (Rasmussen, 1979, 1985; Vicente, 1999). Cells at higher levels of abstraction and decomposition are less detailed than those at lower levels. Therefore, depending on their task demands, actors can zoom in to the details of fine-grained, physical models or zoom out so that these details are aggregated into coarse, purposive models. This mechanism offers actors a way of dealing with systems that would be unmanageable if they always had to observe the whole system in full detail.

Virtually all types of hierarchies, in some sense, represent systems at different levels of detail (Vicente, 1999). The abstraction–decomposition space is unique in that one of its dimensions, the abstraction dimension, is characterized by structural means–ends relations. As a result, actors can decide which ends are relevant in any given situation and, subsequently, concentrate on those means that are pertinent to the desired ends while ignoring other portions of the model. The abstraction hierarchy, therefore, constrains problem solving or reasoning in a way that is purpose oriented.

Previously in this chapter, a distinction was made between structural means–ends and action means–ends relations. Action means–ends hierarchies also may be described as purpose oriented because they identify tasks or actions for achieving specific goals. However, the goals that are appropriate, and thus the tasks or actions for attaining those goals, can only be defined for situations that are known or can be anticipated (Vicente, 1999). Hence, unlike the abstraction hierarchy, action means–ends hierarchies cannot support actors in problem solving during novel or unanticipated events.

Causal versus Intentional Systems

Complex sociotechnical systems may be classified as causal or intentional (Rasmussen et al., 1994). In developing a work domain model of a system, defining whether the system is causal or intentional is important. This step allows analysts to assess the primary constraints on actors in the system and, therefore, the nature of the constraints that need to be represented in the model (Hajdukiewicz, Burns, Vicente, and Eggleston, 1999; Rasmussen, 1999; Rasmussen and Pejtersen, 1995). This section, which explains these points further, may help readers to appreciate why some of the work domain models in this book look so different from one another.

Whether a complex sociotechnical system is categorized as causal or intentional depends on which of two classes of constraints primarily shape actors' behavior in the system. *Causal constraints* have their basis in physical or natural laws (e.g., law of gravity). Such constraints may be regarded as "hard" constraints because they cannot be violated (Burns, Bryant, and Chalmers, 2005, p. 607). *Intentional constraints*, in contrast, stem from social laws, conventions, or values (e.g., company regulations). These constraints may be viewed as "soft" constraints because they can be violated, although it would be socially unacceptable to do so (Burns et al., 2005, p. 607).

Most complex sociotechnical systems have both causal and intentional constraints. A system's classification, therefore, depends on the degree to which actors' behavior is influenced by causal constraints *relative* to intentional constraints. In causal systems, constraints of a causal nature have more bearing on actors' behavior. In intentional systems, actors' behavior is determined mainly by intentional constraints.

Rasmussen et al. (1994) describe a taxonomy for characterizing systems as causal or intentional. Figure 2.13 presents a modified version of this taxonomy, following Hajdukiewicz et al. (1999). This figure has three parts. At the top of the figure are two *coupled* axes, which represent the degree of intentional relative to causal constraints that shapes actors' behavior. The two axes define the *causal–intentional continuum*.

The ovals in the figure denote three categories of systems that occupy different positions along the causal–intentional continuum. These categories are tightly coupled causal systems, loosely coupled intentional systems, and user-driven intentional systems. Within the ovals are some examples of systems that belong in those categories. These systems are presented along the continuum in a way that does not necessarily depict their exact locations but shows their approximate placements and positions relative to each other.

The text at the bottom of the figure provides a simplified description of the nature of the intentional and causal constraints in the systems listed within the ovals.

A system's location on the causal–intentional continuum signifies which of the two classes of constraints has more influence on the behavior of actors. In systems that fall toward the left of the continuum, *tightly coupled causal systems*, a high degree of causal constraints compared with intentional constraints shapes actors' behavior (see the axes at the top of Figure 2.13). This means that their behavior is determined primarily by causal constraints. In systems that fall on the right end of the continuum, *user-driven intentional systems*, intentional constraints have more bearing than causal constraints on actors, which means that their behavior is governed mainly by intentional constraints. In systems that fall toward the right of the middle of the continuum, *loosely coupled intentional systems*, causal constraints play a greater role in shaping actors' behavior than such constraints do in systems located at the extreme right. However, as in user-driven systems, intentional

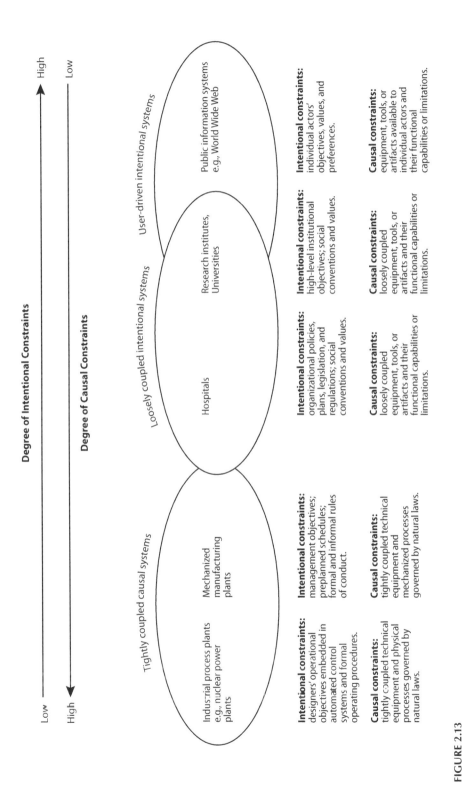

Degree of Intentional Constraints

Low — High

Degree of Causal Constraints

High — Low

Tightly coupled causal systems

Loosely coupled intentional systems

User-driven intentional systems

Industrial process plants e.g., nuclear power plants

Mechanized manufacturing plants

Hospitals

Research institutes, Universities

Public information systems e.g., World Wide Web

Intentional constraints: designers' operational objectives embedded in automated control systems and formal operating procedures.

Causal constraints: tightly coupled technical equipment and physical processes governed by natural laws.

Intentional constraints: management objectives; preplanned schedules; formal and informal rules of conduct.

Causal constraints: tightly coupled technical equipment and mechanized processes governed by natural laws.

Intentional constraints: organizational policies, plans, legislation, and regulations; social conventions and values.

Causal constraints: loosely coupled equipment, tools, or artifacts and their functional capabilities or limitations.

Intentional constraints: high-level institutional objectives; social conventions and values.

Causal constraints: loosely coupled equipment, tools, or artifacts and their functional capabilities or limitations.

Intentional constraints: individual actors' objectives, values, and preferences.

Causal constraints: equipment, tools, or artifacts available to individual actors and their functional capabilities or limitations.

FIGURE 2.13
Taxonomy for characterizing systems as causal or intentional.

constraints carry more weight than causal constraints in loosely coupled systems, so that actors' behavior is controlled primarily by intentional constraints. These three categories of systems are described in more detail shortly.

Given that a system's location on the causal–intentional continuum signifies whether intentional or causal constraints are the primary influence on actors, it also indicates the nature of the constraints that should be represented in a work domain model. Figure 2.14 illustrates this point. Along the horizontal axis are examples of systems that fall at different points along the continuum; these systems are the same as those in Figure 2.13. The vertical axis denotes the abstraction dimension of a work domain model. Accordingly, the figure shows the nature of the constraints that should be represented at purposive (higher) and physical (lower) levels of abstraction given a system's position on the continuum.

The preceding points are now explained further for each of the three categories of systems. Tightly coupled causal systems, such as industrial process plants and mechanized manufacturing plants, are highly structured (Rasmussen et al., 1994). In industrial process plants, for example, intentional constraints stem from designers' operational objectives, which are embedded in automated control systems and formal operating procedures (Figure 2.13). Causal constraints have their basis in tightly coupled technical equipment and physical processes, which are governed by natural laws. Production in these systems (e.g., production of nuclear power) depends heavily on the functions of the technical equipment. The automated control systems and formal operating procedures maintain the plant's performance in accordance with designers' operational objectives, which are fairly stable. The main role of workers is to monitor the functioning of the technical system to ensure that it is consistent with these objectives. This means that workers' activities are driven or paced by the technical system, and individual workers' objectives, values, or preferences are of little significance. Workers' behavior, therefore, is shaped by a high degree of causal constraints relative to intentional constraints. A work domain model of such a system will be dominated by causal constraints at both higher and lower levels of abstraction (Figure 2.14). Intentional constraints may be represented at higher levels. An example of a model of a tightly coupled causal system is that for DURESS II (Bisantz and Vicente, 1994), a thermalhydraulic microworld simulation (Figure 2.15).

User-driven intentional systems, such as public information systems (e.g., the World Wide Web), are characterized by autonomous, casual actors 'working' in their natural environment (Rasmussen et al., 1994). The main sources of intentional constraints in these systems are individual actors' objectives, values, and preferences (Figure 2.13). Causal constraints originate in the equipment, tools, or artifacts available to individual actors and the functional capabilities or limitations of those objects, although the technical component of systems in this category may not be as extensive as those in the two other categories. 'Production' depends heavily on the actors. They select the problem to focus on at any point in time, and they generally have significant control over the pace of their activities. This means that a greater degree of intentional constraints compared with causal constraints controls their behavior. Rasmussen et al. (1994) observe that it is difficult to develop a work domain model that is universally relevant for actors in this type of system, except in very general terms. Intentional constraints may be represented mainly at higher levels of abstraction, and causal constraints may be represented solely at lower levels (Figure 2.14). An example of a model of a user-driven intentional system is that provided by Hansen, Løvborg, and Rasmussen (1991) for a gymnastics computer game (Figure 2.16).

Purposive (higher) levels of abstraction	**Intentional constraints:** designers' operational objectives embedded in automated control systems and formal operating procedures. **Causal constraints:** natural laws.	**Intentional constraints:** management objectives; preplanned schedules; formal and informal rules of conduct. **Causal constraints:** natural laws.	**Intentional constraints:** organizational policies, plans, legislation, and regulations; social conventions and values.	**Intentional constraints:** high-level institutional objectives; social conventions and values.	**Intentional constraints:** individual actors' objectives, values, and preferences.
Abstraction dimension ↕					
Physical (lower) levels of abstraction	**Causal constraints:** tightly coupled technical equipment and physical processes.	**Causal constraints:** tightly coupled technical equipment and mechanized processes.	**Causal constraints:** loosely coupled equipment, tools, or artifacts and their functional capabilities or limitations.	**Causal constraints:** loosely coupled equipment, tools, or artifacts and their functional capabilities or limitations.	**Causal constraints:** equipment, tools, or artifacts available to individual actors and their functional capabilities or limitations.
	Tightly coupled causal systems		Loosely coupled intentional systems		User-driven intentional systems
	Industrial process plants e.g., nuclear power plants	Mechanized manufacturing plants	Hospitals	Research institutes, Universities	Public information systems e.g., World Wide Web

FIGURE 2.14

The nature of the constraints that should be represented in a work domain model given a system's location on the causal–intentional continuum.

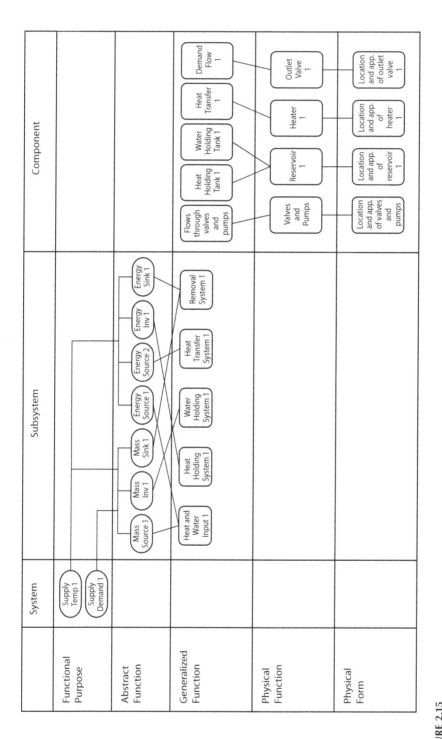

FIGURE 2.15

Abstraction–decomposition space of DURESS II, a tightly coupled causal system. (From Bisantz, A. M., and Vicente, K. J., 1994, Making the abstraction hierarchy concrete, *International Journal of Human-Computer Studies, 40,* 83–117. Copyright 1994. With permission from Elsevier; as adapted by Vicente, 1999.)

Value Systems and Structures	Learn Have fun Score points Competition Task Mastery
Abstract Function, Priority Measures	The "hidden" rules of the game, i.e., priority criteria
General Functions	Playing the game in terms of jump, rotate, land, etc.
Physiological Functions	Joystick movement in terms of left, forward, button press, etc.
Physical Form	Hardware (Screen, stick, computer)

FIGURE 2.16
Abstraction hierarchy of a gymnastics computer game, a user-driven intentional system. (Adapted from Hansen, J. P., Løvborg, L., and Rasmussen, J., 1991, Simulation of cognitive behaviour in computer games, *Proceedings of the Second MOHAWC Workshop*, Risø National Laboratory, Roskilde, Denmark. With permission.)

Loosely coupled intentional systems, such as hospitals, research institutes, and universities, are much less structured than tightly coupled causal systems (Rasmussen et al., 1994). Intentional constraints may be found in organizational policies, plans, legislation, and regulations; high-level institutional objectives; and social conventions and values (Figure 2.13). Causal constraints arise from equipment, tools, or artifacts and the functional capabilities or limitations of those objects, which are loosely coupled rather than tightly coupled. Production (e.g., treatment of patients, publication of research papers) depends heavily on workers. The workers have significant flexibility in establishing goals or priorities and in organizing their daily activities to suit local contingencies or their individual objectives, values, and preferences. In addition, they have considerable control over how physical resources are employed. Accordingly, workers' behavior is governed by a high degree of intentional constraints compared with causal constraints. Work domain models of such systems will be characterized by intentional constraints mainly at higher levels of abstraction and causal constraints entirely at lower levels (Figure 2.14). Rasmussen et al. (1994) provide an example of a model of a loosely coupled intentional system, specifically a health care system (Figure 2.17).

It is worth noting that a system's location on the causal–intentional continuum is highly dependent on how the boundaries of a work domain analysis are defined. The boundaries are influenced by the purpose of the analysis. Different purposes may change the boundaries and, in turn, a system's position on the continuum. Accordingly, the class of constraints that is the principal determinant of actors' behavior in the system, and the nature of the constraints that should be represented in a model, may vary. For example, a home may be classified as a tightly coupled causal system if the purpose of the analysis is to design an ecological interface for a tool to assist builders with the physical construction of a house. Alternatively, the same home may be classified as a loosely coupled intentional system if the purpose of the analysis is to design a decision support tool to assist inhabitants with meal planning. This topic, including the examples, is discussed more completely in Chapter 8, which presents guidelines for work domain analysis.

Goals and Constraints	Cure patient; Research, Training MDs; Public opinion; Legal, economic, and ethical constraints
Priority Measures, Flow of Values and Material	Categories of diseases: Cost of treatments, patient suffering, research relevance
General Functions and Activities	Cure, diagnostics, surgery, medication, etc. Research, clinical, experiments
Physical Activities in Work, Physical Processes of Equipment	Specific research and treatment procedures; Use of tools and equipment
Appearance, Location and Configuration of Material Objects	Material resources, patients, personnel, equipment; Medicine, tools, etc.

FIGURE 2.17

Part of an abstraction hierarchy of a health care system, a loosely coupled intentional system. The full model is shown in Figure 6.3. (Adapted from Rasmussen, J., Pejtersen, A. M., and Goodstein, L. P., 1994, *Cognitive systems engineering*, Wiley, New York. Copyright © 1994 John Wiley & Sons. This material is reproduced with permission of John Wiley & Sons, Inc.)

Formats for Work Domain Models

A work domain model may be presented in different formats. This section provides a brief overview of four formats that analysts commonly adopt, which I refer to as the graphical abstraction–decomposition space, the graphical abstraction hierarchy, the tabular abstraction–decomposition space, and the tabular abstraction hierarchy.

As the following discussion shows, the main differences between these formats are whether they are tabular or graphical, whether they indicate the level of decomposition of the constraints in the model, and whether they portray means–ends relationships between those constraints. The format that analysts choose may depend on their individual preferences or the intended application of the model. Analysts may adopt multiple formats in order to emphasize different aspects of a work domain.

The four formats are depicted generically to highlight their similarities and differences. In addition, to assist in the comparison, all of the formats are shown with five levels of abstraction and three levels of decomposition. This review may help readers to understand the variations in the appearance of the work domain models in this book.

Graphical Abstraction–Decomposition Space

Figure 2.18a portrays a graphical abstraction–decomposition space; the nodes within the cells denote constraints. This format shows both the levels of abstraction and the levels of decomposition of the constraints in the model, as well as the means–ends relation-

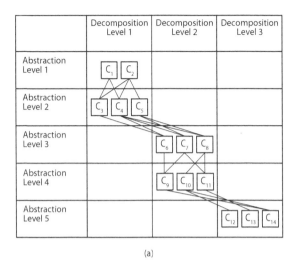

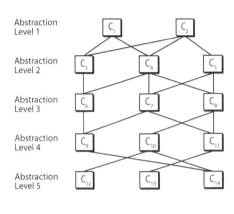

(a)

(b)

(c)

(d)

FIGURE 2.18

Four formats for presenting work domain models: (a) graphical abstraction–decomposition space; (b) graphical abstraction hierarchy; (c) tabular abstraction–decomposition space; (d) tabular abstraction hierarchy. C, constraint.

ships between those constraints. The model of DURESS II (Bisantz and Vicente, 1994), a thermalhydraulic microworld simulation, in Figure 2.15 is in this format.

Graphical Abstraction Hierarchy

Figure 2.18b depicts a graphical abstraction hierarchy. This format is similar to the graphical abstraction–decomposition space (Figure 2.18a) in that it shows the level of abstraction of the constraints and their means–ends relationships. However, it does not show the level of decomposition to which the constraints belong. A model of the game of baseball (Vicente and Wang, 1998) in the next chapter (see Figure 3.10) is in this format.

There are many possibilities for the level of decomposition of the constraints in the graphical abstraction hierarchy. Consider the graphical abstraction–decomposition space in Figure 2.18a. This figure demonstrates one possibility. Other possibilities are that the constraints belong to a different combination of levels of decomposition or solely to the first, second, or third level. Unless it is stated explicitly, it is difficult to know with certainty at which level of decomposition the constraints in the graphical abstraction hierarchy fall.

Tabular Abstraction–Decomposition Space

Figure 2.18c portrays a tabular abstraction–decomposition space; the entries within the cells represent constraints. Like the graphical abstraction–decomposition space (Figure 2.18a), this format shows both the levels of abstraction and the levels of decomposition of the constraints. However, unlike the two graphical formats (Figures 2.18a and 2.18b), this format does not show the means–ends relationships between those constraints. The model of a home presented previously in this chapter (Figure 2.7) is in this format.

Tabular Abstraction Hierarchy

Figure 2.18d depicts a tabular abstraction hierarchy. Consistent with all of the preceding formats, this format shows the level of abstraction of the constraints. However, it does not show their level of decomposition or means–ends relationships. The models of a gymnastics computer game (Hansen et al., 1991; Figure 2.16) and health care system (Rasmussen et al., 1994; Figure 2.17) are in this format.

Summary

Work domain analysis is concerned with modeling the functional structure of the physical, social, or cultural environment of actors in a system. A work domain model is useful because it

1. identifies the constraints on actors;
2. reveals the possibilities for action available to them;
3. offers a rationale for their behavior;
4. remains relatively constant across a range of situations;
5. leads to designs that support actors in handling a variety of events, including those that are novel or unforeseen.

The main modeling tools for work domain analysis are the abstraction–decomposition space and the abstraction hierarchy. The key features of the abstraction–decomposition space, which subsumes the abstraction hierarchy, are the abstraction and decomposition dimensions, structural means–ends and part–whole relations, and constraints. These features, illustrated in this chapter with a model of a home (Figure 2.7), make the abstraction–decomposition space a tool for modeling work domains, not actors' behavior. This tool represents work domains in a way that is compatible with the characteristics of human problem solving or reasoning.

Complex sociotechnical systems may be categorized as causal or intentional. This classification is important because it allows analysts to assess the primary constraints on actors in a system and, accordingly, the nature of the constraints that need to be represented in a model.

A work domain model may be presented in different formats. Four formats that analysts commonly adopt are the graphical abstraction–decomposition space, the graphical abstraction hierarchy, the tabular abstraction–decomposition space, and the tabular abstraction hierarchy.

In the next part of this book, Chapters 3 to 7, I attend to the concepts of work domain analysis in more depth, focusing on topics that were not dealt with comprehensively in earlier texts. These chapters are organized largely around the abstraction–decomposition space.

Section II

Concepts

3

Abstraction

OVERVIEW   In creating a work domain model of a system, it is useful for analysts to think about the number, types, and labels of the levels of abstraction that should be included in the model. This chapter reviews representations of a range of systems to establish whether a widely accepted set of levels of abstraction, that described by Rasmussen (1986) and Rasmussen et al. (1994), can be confidently adopted for modeling any system or whether it may be prudent for analysts not to assume this standard set in every case. Subsequently, detailed descriptions are provided of the standard set as it is clearly useful in numerous, although not necessarily all, contexts. These descriptions are illustrated with examples from work domain models of a home, a library, a military system, and a pasteurization plant.

Number, Types, and Labels of Levels of Abstraction

The process of developing a work domain model of a system involves contemplating the number, types, and labels of the levels of abstraction to include in the model. One of the goals of analysts in making these decisions is to construct a valid or faithful representation of the system. Another is to produce a model that is suitable for its intended application. Work domain models that do not address these concerns effectively may result in poor designs or solutions, which do not foster system safety, productivity, or workers' health.

Since Rasmussen (1979, 1985) formulated the abstraction–decomposition space, this tool has been used to develop work domain models of a variety of systems. One question worth examining now, given the range of existing models, is whether the conventional set of levels of abstraction described by Rasmussen (1986) and Rasmussen et al. (1994) is suitable for modeling every system. If so, analysts may model any system using this standard set. Otherwise, analysts may choose not to assume this set in every case.

As this topic is potentially a controversial one, this section reviews at length the work domain models of an array of systems to address this question. Readers who are not interested in the details of this review may proceed directly to the final two paragraphs of this section, which provide a summary of the main findings.

Originally, Rasmussen (1979, 1985) developed a set of five levels of abstraction for modeling process control and computer maintenance systems. Figure 3.1, which recasts a figure in Rasmussen's (1986) book, summarizes the labels and the concepts that characterize this set. The labels vary slightly throughout Rasmussen's publications, including within a text, suggesting that he is not prescriptive about them. (For example, compare the sets of labels in Figures 3.1 and 3.2, which both appear in Rasmussen's book.) Nevertheless, those in Figure 3.1—functional purpose, abstract function, generalized functions, physical functions, and physical form—are typical of this set.

Labels	Concepts
Functional Purpose	Production flow models, system objectives, constraints, etc.
Abstract Function	Causal structure: mass, energy, and information flow topology, etc.
Generalized Functions	"Standard" functions and processes: feedback loops, heat transfer, etc.
Physical Functions	Electrical, mechanical, chemical processes of components and equipment
Physical Form	Physical appearance and anatomy; material and form; locations, etc.

FIGURE 3.1

Rasmussen's (1986) summary of a set of five levels of abstraction. (This summary was also published in an earlier paper. Adapted from Rasmussen, J., 1983, Skills, rules, and knowledge; signals, signs, and symbols, and other distinctions in human performance models, *IEEE Transactions on Systems, Man, and Cybernetics, 13*(3), 257–266. © 1983 IEEE. With permission; incorporating modifications made by Rasmussen, 1986.)

To illustrate this set, Rasmussen (1986) presents an abstraction hierarchy of a manufacturing system (Figure 3.2). This suggests that he held the view that the five levels of abstraction he had developed originally for modeling process control and computer maintenance systems could be extended to manufacturing systems. Even so, he states explicitly that the number and the types of levels of abstraction that are included in a work domain model depend on the kind of system and the purpose of the analysis.

Subsequently, Rasmussen et al. (1994) provided another detailed description of a set of five levels of abstraction. Figure 3.3, which replicates a figure in that book, presents the labels and the concepts that define that set. Both the labels (i.e., purposes and constraints, abstract functions and priority measures, general functions, physical processes and activities, and physical form and configuration) and the descriptions of the concepts appear relatively different from those Rasmussen (1986) presented earlier (Figure 3.1). However, there is no implication by Rasmussen et al. that their set of five levels of abstraction is different from Rasmussen's earlier set or, in other words, different from the set that had been developed for modeling process control and computer maintenance systems. On the basis of this point, as well as the summaries (Figures 3.1 and 3.3) and the discussions of the five levels of abstraction in the two texts, it may be argued that the differences between the sets are mainly in language and not in the nature of the concepts that characterize each level of abstraction.

Other evidence that the sets of five levels of abstraction presented by Rasmussen (1986) and Rasmussen et al. (1994) are the same revolves around two facts. First, Rasmussen and his colleagues have published summaries that combine the labels from one set with the descriptions of concepts in the other. For example, Rasmussen's 1988 summary, shown in Figure 3.4, describes the concepts in much the same way as Rasmussen et al. did in 1994 (Figure 3.3) but uses many of the labels from Rasmussen's 1986 work (Figure 3.1).

Second, Rasmussen and his colleagues present work domain models of systems with labels similar to those in Rasmussen et al.'s 1994 work (Figure 3.3), despite the fact that such

Purpose	Market relations Supply sources Energy and waste constraints Safety requirements
Abstract Function	Flow of energy and mass, products, monetary values Mass, energy balances Information flow structure in system and organization
Generic Function	Production, assembly, maintenance Heat removal, combustion, power supply Feedback loops
Physical Function	Physical functioning of equipment and machinery Equipment specifications and characteristics Office and workshop activities
Physical Form	Form, weight, color of parts and components Their location and anatomical relation Building layout and appearance

FIGURE 3.2

Abstraction hierarchy of a manufacturing system that Rasmussen (1986) used to illustrate the set of five levels of abstraction in his book. (Adapted from Rasmussen, J., 1986, *Information processing and human-machine interaction: An approach to cognitive engineering*, North-Holland, New York. Copyright © 1986 North-Holland. With permission from Elsevier.)

systems were portrayed originally with the types of labels that Rasmussen presented in 1986 (Figure 3.1). Rasmussen (1988), for instance, uses labels of the kind in Rasmussen et al.'s 1994 work in an abstraction hierarchy of a process control system (Figure 3.5). Likewise, Rasmussen, Pejtersen, and Schmidt (1990) adopt such labels in an abstraction hierarchy of a manufacturing system (Figure 3.6). This model of a manufacturing system may be compared directly with another in Figure 3.2, which has labels like those in Rasmussen's 1986 work. The main difference between these representations seems to be that the manufacturing system in Figure 3.6 is modeled more like an intentional system, whereas the one in Figure 3.2 is modeled more like a causal one. The types of concepts that characterize each level of abstraction in the two models are the same.

If we accept that the sets of five levels of abstraction in the texts by Rasmussen (1986) and Rasmussen et al. (1994) are the same, then judging from the various models produced by Rasmussen and his colleagues, it seems that the original set, which was developed in the context of process control and computer maintenance systems, is suitable for modeling several other types of systems. These include manufacturing systems (Rasmussen, 1986; Rasmussen et al., 1990); health care systems (Rasmussen et al., 1990, 1994); cities (Rasmussen et al., 1994); libraries (Rasmussen et al., 1994); offices (Rasmussen, 1988); computer games (Hansen et al., 1991); emergency management systems (Rasmussen, 1988; Rasmussen, Pedersen, and Grønberg, 1987); military systems (Rasmussen, 1998); and engineering design systems (Rasmussen, 1988, 1990). The evolution in the language of the labels and descriptions of the levels most likely reflects an attempt to accommodate this broader range of systems, which includes intentional as well as causal systems. This conclusion may be justified in light of Reising's (2000) commentary on his dialogues with Rasmussen, which is discussed later in this section.

Labels	Concepts
Purposes and Constraints	Properties necessary and sufficient to establish relations between the performance of the system and the reasons for its design, that is, the purposes and constraints of its coupling to the environment. *Categories are in terms referring to properties of environment.*
Abstract Functions and Priority Measures	Properties necessary and sufficient to establish priorities according to the intention behind design and operation: Topology of flow and accumulation of mass, energy, information, people, monetary value. *Categories in abstract terms, referring neither to system nor environment.*
General Functions	Properties necessary and sufficient to identify the 'functions' which are to be coordinated irrespective of their underlying physical processes. *Categories according to recurrent, familiar input-output relationships.*
Physical Processes and Activities	Properties necessary and sufficient for control of physical work activities and use of equipment: To adjust operation to match specifications or limits; to predict response to control actions; to maintain and repair equipment. *Categories according to underlying physical processes and equipment.*
Physical Form and Configuration	Properties necessary and sufficient for classification, identification and recognition of particular material objects and their configuration: for navigation in the system. *Categories in terms of objects, their appearance and location.*

FIGURE 3.3

Rasmussen et al.'s (1994) summary of a set of five levels of abstraction. (Adapted from Rasmussen, J., Pejtersen, A. M., and Goodstein, L. P., 1994, *Cognitive systems engineering*, Wiley, New York. Copyright © 1994 John Wiley & Sons. This material is reproduced with permission of John Wiley & Sons, Inc.)

Labels	Concepts
Purpose, Constraints	Properties necessary and sufficient for relating the performance of the system with the reasons for design, with requirements of environment. *Categorization in terms referring to properties of the environment.*
Abstract Function	Properties necessary and sufficient to establish relationships according to design or intention: energy, value, information, truth, etc. Relationship to underlying causal structure and function is depending on convention and design choice. *Categorization in abstract terms, referring neither to system nor to the environment.*
Generalized Function	Properties necessary and sufficient to establish "black box" input-output models of functions irrespective of underlying implementation; this level is necessary for coordination of different physical processes to serve joint higher level purpose. *Categorization according to recurrent, familiar input-output relationships.*
Physical Function	Properties necessary and sufficient for use of object: for adjustment of object for use, to adjust to limits of use, to predict whether objects will serve particular use, to select part to move for control of physical process. *Categorization according to underlying physical process.*
Physical Form	Properties necessary and sufficient for classification and recognition of material objects.

FIGURE 3.4

Summary of a set of five levels of abstraction in Rasmussen (1988). (Adapted from Rasmussen, J., 1988, A cognitive engineering approach to the modeling of decision making and its organization in: Process control, emergency management, CAD/CAM, office systems, and library systems. In W. B. Rouse, Ed., *Advances in man-machine systems research*, 4, JAI Press, Greenwich, CT, 165–243. Copyright © 1988 JAI Press. With permission from Elsevier.)

Goals and Values, Constraints	Customer and Market Relations; Competitors; Production Volume Requirements; Legal Requirements for Financial Relations and Environmental Protection; Work Safety Legislation; Agreements with Workers' Unions
Flow, Distribution, and Accumulation of Material, Energy, Monetary Values, and Manpower	Topology of Product Flow and of Major Mass, Energy, and Information Flows; Major Mass and Energy Balance Systems, their Properties and Limitations; Flow of Monetary Values; Manpower Turnover; etc.
General Functions and Activities	Production Functions; Cooling, Heating, Purification Functions; Control Functions and Feedback Loops, etc.
Specific Work Processes, Physical Processes of Equipment	Technical Characteristics of Equipment, Machinery, and Components; Their Capabilities and Limitations, Control Characteristics; Content of Manuals and Technical Specifications; Maintenance Properties
Appearance, Location, and Configuration of Material Resources	Material Characteristics, Sizes, Weight, Appearance, and Location; Anatomy and Configuration of Equipment and Installations; Building Layout; Drawings; Access Roads and Site Topography

FIGURE 3.5

Abstraction hierarchy of a process control system with labels like those in Rasmussen et al. (1994). (Adapted from Rasmussen, J., 1988, A cognitive engineering approach to the modeling of decision making and its organization in: Process control, emergency management, CAD/CAM, office systems, and library systems. In W. B. Rouse, Ed., *Advances in man-machine systems research*, 4, JAI Press, Greenwich, CT, 165–243. Copyright © 1988 JAI Press. With permission from Elsevier.)

The range of systems to which Rasmussen's (1986) and Rasmussen et al.'s (1994) set of five levels of abstraction is applicable may be investigated further by examining the work domain models of other analysts. Table 3.1 lists some of the types of systems that various analysts, including Rasmussen and his colleagues, have modeled.

Of the models created by other analysts, the majority—by far—have five levels of abstraction. The labels of these levels are usually the same as those in the texts by Rasmussen (1986; Figure 3.1) or Rasmussen et al. (1994; Figure 3.3) or are variations or combinations of those labels. The concepts at each level also appear to be mainly the same as those described by Rasmussen and Rasmussen et al. In fact, analysts often cite these publications as the sources of the levels of abstraction in their models. For these reasons, from this point, the set of five levels of abstraction in the texts by Rasmussen and Rasmussen et al. is referred to in this book as the 'standard set' or 'standard levels' unless citing these publications explicitly is necessary for clarity.

Goals and Constraints	Finance Policy: Manufacturing, Bonds and Investment. Social Company Policies. Market Policy: Custom-Design; High-Tech; Public Image: Innovative, Advanced-Tech. Consumer-Population: Age, Education; Location, etc. Supplier-population: Policy, Reliability, Research Status. Tax and Business Laws. Regulations of Branch Associations. Union Agreements. Workers Protection Regulations			
Priority Measures, Flow of Values and Material	Flow of Funds from Sources: Sales, Bonds, Interest, etc. to Sinks: Purchasing, Salaries, Investment, Losses. Flow of Material from Sources: Suppliers to Departments to Products to Sinks: Markets, Customers, and Waste. Flow of Equipment & Tools: Supplier, Purchasing, Installation, Maintenance, Modification, Loss and Rejection. Flow of Manpower: Education, Company Training, Department, Function, Promotion, Retirement, Lay-off			
General Functions and Activities	Research and Development: Follow State of the Art; Conceive new Products, Test Ideas	Manufacturing: Prepare for Production; Plan; Program Machines; Train Staff; Control and Monitor; Test and Pack	Purchasing & Sales: Accounting; Evaluate Suppliers; Predict Consumption; Market Evaluation; Production Schedules; Pricing	Personnel Administration: Hiring, Laying-off; Fixing Salaries, Evaluation; Planning Work Schedules and Shifts
Physical Activities in Work, Physical Processes of Equipment	Analyze Competing Products, Literature, Patents. Experiments; Prototyping	Processes of Equipment and Tools; Planning Processes and Methods	Accounting and Planning Tools; Spreadsheets & Budgetary Tools; Market & Sales Reviews; Statistics	Interviews; Reading Advertisements; Negotiations; Statistics on Performance and Expenses; Compare and Promote
Appearance, Location and Configuration of Material Objects	Library Items; Drawings; Equipment; Specifications and Location	Location and Specifications of Machinery, Buildings, Components, etc.	Suppliers and Customers, Identification, Location; Product and Raw Material Specifications and Inventories	Personnel Identification and Characteristics, Salary Lists; Addresses, Family Information; Transportation Information; Vacancy Lists

FIGURE 3.6
Abstraction hierarchy of a manufacturing system with labels similar to those in Rasmussen et al. (1994). (Adapted from Rasmussen, J., Pejtersen, A. M., and Schmidt, K., 1990, *Taxonomy for cognitive work analysis* (Risø-M-2871), Risø National Laboratory, Roskilde, Denmark. With permission.)

Models with fewer than five levels of abstraction typically comprise a subset of the standard levels and, therefore, do not signify fundamental departures from the standard set. For instance, Burns and Vicente (1995) present a model of an engineering design system with three levels of abstraction.* The labels of these levels, which are objectives, processes, and physical components, and the concepts that characterize these levels are akin to a

---

* See Figure 6.1 for the model of the engineering design system.

TABLE 3.1

Some Types of Systems for Which Analysts Have Produced Work Domain Models

Types of Systems	Examples of Publications
Process control systems	Bisantz and Vicente (1994)
	Burns (2000)
	Dinadis and Vicente (1996)
	Ham, Yoon, and Han (2008)
	Itoh, Sakuma, and Monta (1995)
	Jamieson and Vicente (2001)
	Lau, Jamieson, Skraaning, and Burns (2008)
	Reising and Sanderson (2002b)
Manufacturing systems	Kinsley, Sharit, and Vicente (1994)
	Rasmussen (1986)
	Rasmussen et al. (1990)
Medical systems	Hajdukiewicz, Vicente, Doyle, Milgram, and Burns (2001)
	Sharp (1996)
	Thompson, Hickson, and Burns (2003)
	Watson and Sanderson (2007)
	Rasmussen et al. (1990, 1994)
Military systems	Bisantz, Roth, Brickman, Gosbee, Hettinger, and McKinney (2003)
	Burns et al. (2005)
	Hajdukiewicz et al. (1999)
	Jenkins, Stanton, Salmon, Walker, and Young (2008)
	Lintern (2006)
	Naikar, Pearce, Drumm, and Sanderson (2003)
	Naikar and Sanderson (1999, 2001)
	Torenvliet, Jamieson, and Chow (2008)
	Rasmussen (1998)
Aviation systems	Ahlstrom (2005)
	Amelink, Mulder, van Paassen, and Flach (2005)
	Borst, Suijkerbuijk, Mulder, and van Paassen (2006)
	Dinadis and Vicente (1999)
	Ellerbroek, Visser, van Dam, Mulder, and van Paassen (2009)
	Van Dam, Mulder, and van Paassen (2008)
	Xu (2007)
Maritime systems	Morel and Chauvin (2006)
Gaming systems	Burns and Proulx (2002)
	Vicente and Wang (1998)
	Hansen et al. (1991)
Engineering design systems	Benda and Sanderson (1998)
	Burns and Vicente (1995, 2000)
	Dainoff and Mark (1995)
	Pejtersen, Sonnenwald, Buur, Govindaraj, and Vicente (1997)
	Xu, Dainoff, and Mark (1999)
	Rasmussen (1988, 1990)
Information management systems	Albrechtsen and Pejtersen (2003)
	Simons, Dainoff, and Mark (2007)

TABLE 3.1 (*Continued*)

Some Types of Systems for Which Analysts Have Produced Work Domain Models

Types of Systems	Examples of Publications
Road transport or driving systems	Jenkins, Stanton, Walker, and Young (2007)
	Stoner, Wiese, and Lee (2003)
Educational systems	Dorneich (2002)
	Fidel and Pejtersen (2004)
Emergency management systems	Hajdukiewicz et al. (1999)
	Moray, Sanderson, and Vicente (1992)
	Rasmussen (1988)
	Rasmussen et al. (1987)
Network management systems	Burns, Kuo, and Ng (2003)
	Duez and Vicente (2005)
Electronic devices	Jenkins, Stanton, Walker, Salmon, and Young (2010)
	Mazaeva and Bisantz (2007)
Cities	Rasmussen et al. (1994)
Libraries	Rasmussen et al. (1994)
Offices	Rasmussen (1988)

subset of the standard levels. Furthermore, Burns and Vicente (2000) acknowledge that a finer gradation of levels of abstraction may have been valid for modeling the engineering design system, but that the three levels in their representation were sufficient for their design problem.

A few work domain models have four levels of abstraction. One example is a model of the natural environment,[*] which is part of a representation of a naval command-and-control system, specifically a frigate (Burns et al., 2005). The labels (abstract function, generalized function, physical function, and physical form) and the concepts that define the levels of abstraction in this model are comparable to those for the last four levels in the standard set, although the language that Rasmussen (1986) and Rasmussen et al. (1994) use to describe the concepts of that set is better suited to engineered systems than to natural systems. According to Burns and Hajdukiewicz (2004), the functional purpose level was not included in the model because the natural environment is not an engineered system and acts without purpose. Nevertheless, they allow that it may be possible to describe this level of abstraction for the natural environment.

Other work domain models with four levels of abstraction do not depict the physical form or physical objects of the systems in question. In the case of several models of computer simulations of various systems, such as a pasteurization plant[†] (Reising and Sanderson, 2002b); a coal-fired power plant (Burns, 2000); and a fluid catalytic cracking unit in a petrochemical plant[‡] (Jamieson and Vicente, 2001), the analysts state that they excluded this level from their representations because it does not exist in the computer simulations. The actual systems, though, do have a physical form. Otherwise, the labels and the concepts that define the four levels of abstraction in these models are the same as those for the first four levels in the standard set.

---

[*] See Figure 6.10 for the model of the natural environment.
[†] See Figure 5.3 for the model of the microworld simulation of a pasteurization plant.
[‡] See Figure 4.1a for the model of the simulation of a fluid catalytic cracking unit in a petrochemical plant.

According to Reising (2000), Rasmussen indicated during discussions between them that while at one stage he thought his original set of five levels of abstraction may have been an artifact of his work with thermodynamic systems, he had since become convinced that this set is conceptually necessary and sufficient. Rasmussen suggested that the five levels could be relabeled functional purposes, value and priority measures, purpose-related functions, object-related processes, and physical objects. Reising notes that these labels make clearer the concepts that characterize each level, the relationships between the concepts at different levels, and the relevance of this set to intentional systems. There seems to be no implication in Rasmussen's discussions with Reising that the new labels signify a change in the concepts relevant to the standard set.

In contrast to Rasmussen's view that his original set of five levels of abstraction is conceptually necessary and sufficient, Vicente (1999) argues that although the same five levels have been useful for modeling a variety of systems, there is no reason to believe that they will be suitable for all systems. According to Vicente, "the number of levels, and especially their content, may vary as a function of the types of constraints in each domain" (p. 164). He points out that Sharp's (1996) and Hajdukiewicz's (1998) models of the human body (Figures 3.7a and 3.8a, respectively) require a different set of five levels.

From inspecting Sharp's (1996) and Hajdukiewicz's (1998) models, it is not clear that the concepts in them are fundamentally different from those in the standard set. However, Sharp mentions explicitly that he found that Rasmussen's (1986) five levels of abstraction were not appropriate for creating a representation of the human body, specifically a model of tissue oxygenation as it relates to neonatal intensive care. This led him to examine the phrases and terminology of the language of clinicians in the hospital where he did his study to derive a set of levels that was appropriate for his model.

Sharp's (1996) descriptions of the concepts in his model are shown in Figure 3.7b. On the basis of these descriptions, as well as the content of his model (Figure 3.7a), it may be argued that Sharp ended up with essentially the same five levels of abstraction as those in the standard set. The main difference appears to be simply that the content of his model and the language he uses to describe the concepts in it are specific to the system he studied. The relevance of Rasmussen's (1986) set of five levels of abstraction to natural systems is sometimes difficult to see because the language he employs for describing these levels is better suited to engineered systems.

The argument that Sharp's (1996) model comprises essentially the same five levels of abstraction as those in the standard set may be further supported in light of Hajdukiewicz's (1998) work. To represent the work domain of an anesthesiologist in an operating room, Hajdukiewicz constructed an abstraction–decomposition space of the human body (Figure 3.8a) as well as one of its cardiovascular system (Figure 3.8b). Hajdukiewicz's model was clearly informed by Sharp's earlier work. The labels of the first three levels of abstraction match those in Sharp's model. In addition, the labels of the last two levels are the same as those in the first iteration of Sharp's model. Hajdukiewicz also states plainly that his representation of the human body is similar to the one developed by Sharp.

Hajdukiewicz (1998), however, does not claim that the five levels of abstraction in his model are different from the standard levels. In fact, in describing his model of the cardiovascular system, he supplies his own labels as well as Rasmussen's (1986) as section titles for each of the levels. Hajdukiewicz also relies on Rasmussen's labels and the style and terminology of Rasmussen et al.'s (1994) descriptions to explain the concepts at each level. For example, to describe the first level, he states that "This level of abstraction represents the functional purposes governing the interaction between the cardiovascular system and the environment" (p. 58).

	Body	Body Systems	Body Organs	Cellular
Purpose	Homeostasis: Maintain internal environment			
Balances	Balance oxygen supply and demand	Oxygen balance in respiratory, Cardio-vascular and Metabolic systems	Oxygen balance in alveoli, pulmonary blood, arterial blood, capillary blood, and tissues	Balance oxygen supply and demand at the mitochondria of all cells in the body to support anaerobic metabolism
Processes	Homeostatic processes	Oxygenation, Ventilation	Alveolar Ventilation, Tissue Perfusion, Circulation	
Transport, Storage and Control			Homeostatic response, Vascular volume, Vasoconstriction, Blood pressure, Blood flow	
Physical Form			X-rays: lungs, heart size & location	

(a)

Labels	Concepts
Purpose	Homeostasis: maintain internal environment
Balances	To maintain the internal environment the supply and demand for nutrients must be balanced
Processes	The process that connects the chambers that are in balance. Process corresponds to regulated flows of oxygen
Transport, Storage and Control	The components that make up the processes, and the storage chambers
Physical Form	The actual arrangements and interconnections of the various body sub-systems

(b)

FIGURE 3.7
(a) Sharp's (1996) abstraction–decomposition space of the human body. (Adapted from Sharp, T.D., 1996, *Progress towards a development methodology for decision support systems for use in time-critical, highly uncertain, and complex environments,* unpublished doctoral dissertation, University of Cincinnati, Ohio. With permission of the author.)
(b) Sharp's (1996) descriptions of the concepts at each level of abstraction in his model.

	Whole Body	System	Organ	Tissue	Cell
Purposes	Homeostasis (Maintenance of Internal Environment)	Adequate Circulation, Blood Volume, Oxygenation, Ventilation	Adequate Organ Perfusion, Blood Flow	Adequate Tissue Oxygenation and Perfusion	
Balances	Balances: Mass and Energy Inflow, Storage, and Outflow *	System Balances: Mass and Energy Inflow, Storage, Outflow, and Transfer *	Organ Balances: Mass and Energy Inflow, Storage, Outflow, and Transfer *	Tissue Balances: Mass and Energy Inflow, Storage, Outflow, and Transfer *	
Processes	Total Volume of Body Fluid, Body Temperature, Supply: O_2, Fluids, Nutrients. Sink: CO_2, Fluids, Wastes	Circulation, Oxygenation, Ventilation, Circulating Volume	Perfusion Pressure, Organ Blood Flow, Vascular Resistance	Tissue Oxygenation, Respiration, Metabolism	Cell Metabolism, Chemical Reaction, Binding, Inflow, Outflow
Physiology			Organ Function	Tissue Function	Cellular Function
Anatomy			Organ Anatomy	Tissue Anatomy	Cellular Anatomy

*Balances include Water, Salt, Electrolytes, pH, O_2, CO_2

(a)

	System	Sub-system	Organ	Component
Purposes	Adequate Circulation and Blood Volume			
Balances	Cardiovascular System: Mass Inflow, Storage, and Outflow	Pulmonary and Systemic Systems: Balance Mass Flows: Mass Inflow, Storage, Outflow, and Transfer	Organ Vascular Network: Balance Mass Flows: Mass Inflow, Storage, Outflow, and Transfer	Vascular Components: Balance Mass Flows: Mass Inflow, Storage, Outflow, and Transfer
Processes	Circulation, Volume, Fluid Supply and Sink	Pulmonary and Systemic Circulation (Pressure, Flow, Resistance) and Volume, Fluid Supply and Sink	Cardiac Output, Organ Circulation (Pressure, Flow, Resistance), Fluid Supply and Sink from each Vascular Network	Circulation through Vascular Components (Pressure, Flow, Resistance), Vascular Blood Volume, Fluid Supply and Sink
Physiology			Cardiac Function (Heart Rate, Rhythm)	Atrial and Ventricular Function, Arterial, Arteriolar, Capillary, Venule, Venous Function
Anatomy				Cardiac and Vascular Anatomy

(b)

FIGURE 3.8
Hajdukiewicz's (1998) abstraction–decomposition spaces of (a) the human body and (b) the cardiovascular system. (Adapted from Hajdukiewicz, J.R., 1998, *Development of a structured approach for patient monitoring in the operating room,* unpublished master's thesis, University of Toronto, Ontario, Canada. With permission of the author.)

Irrespective of whether we agree that the five levels of abstraction in Sharp's (1996) and Hajdukiewicz's (1998) models are essentially the same as those in the standard set, there are two observations that support Vicente's (1999) argument that the same set may not apply to all systems. First, two models of the games of chess (Figure 3.9) and baseball (Figure 3.10), which were developed by Vicente and Wang (1998), arguably represent concepts at two levels, at least, that are different from the standard levels. In particular, the concepts at the

Purpose	Player's goals (e.g., to win, to take risks, and to play an elegant game)
Strategies	Higher level lines of development that can achieve purpose
Tactics	Lower level combinations that can implement a strategy
Paths	Available paths of movement that can be selected to execute a tactic
Board	Spatial location and appearance of pieces on board

FIGURE 3.9
Abstraction hierarchy of the game of chess. (Adapted from Vicente, K. J., and Wang, J. H., 1998, An ecological theory of expertise effects in memory recall, *Psychological Review, 105*(1), 33–57. American Psychological Association, Inc. Copyright 1998. With permission.)

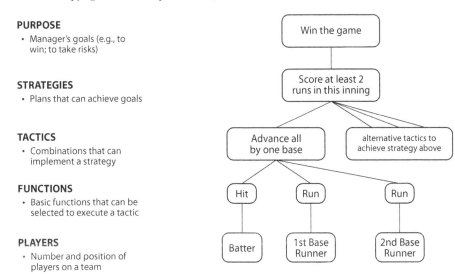

FIGURE 3.10
Abstraction hierarchy of the game of baseball. The right side of the figure depicts a subset of this model that is activated by a vignette described in Vicente and Wang (1998). (From Vicente, K. J., and Wang, J. H., 1998, An ecological theory of expertise effects in memory recall, *Psychological Review, 105*(1), 33–57. American Psychological Association, Inc. Copyright 1998. With permission.)

second (strategies) and third (tactics) levels appear to be unique. It is important to acknowledge that Vicente and Wang state that their models do not reflect comprehensive analyses of the games of chess and baseball. Their representations were created to illustrate the role of the abstraction hierarchy in a theory that explains the effects of expertise on memory recall.

Second, although numerous models across a range of systems have essentially the same five levels of abstraction as those in the standard set, it may be argued that the main reason for this similarity is that analysts were working from example. That is, analysts may not have investigated which levels were most suitable for modeling their systems but, instead, may have simply adopted the standard ones. As a rule, analysts have not claimed to have analyzed the levels to include in their representations independently of those in the standard set. Rather, as indicated previously, they have tended to cite the texts by Rasmussen (1986) and Rasmussen et al. (1994) as the sources of the levels of abstraction in their models.

On the basis of the preceding observations, this book offers two approaches for defining the abstraction dimension of a work domain model. The first is to adopt the standard set of five levels of abstraction presented by Rasmussen (1986) and Rasmussen et al. (1994). This approach recognizes that while analysts may have been working from example, it cannot be denied that this set has been useful for modeling a variety of systems. Furthermore, these models have been beneficial for a range of applications. Therefore, given that it may not be necessary or possible to perform a detailed analysis of the abstraction dimension "from scratch" (Rasmussen et al., 1990, p. 35) on every occasion, the following section provides comprehensive descriptions of the standard levels. In addition, the guidelines for work domain analysis in Chapter 8 include a set of prompts and keywords for defining the constraints to include in a model as a function of the standard levels.

The second approach is to analyze the levels of abstraction to include in a model without assuming the standard set. This approach recognizes that one reason for the considerable similarity between the levels of abstraction in a range of models and the standard ones may be that analysts were working from example. Furthermore, although the standard levels have been useful in a variety of cases, they are not necessarily the best possible set for all systems. If an alternative set offers a more valid or faithful representation of a system, it may lead to better designs or solutions for that system and is, therefore, preferable. The second approach also recognizes that at least two levels in the models of the games of chess and baseball (Vicente and Wang, 1998) are arguably different from the standard levels. For these reasons, the guidelines for work domain analysis in Chapter 8 also suggest a strategy for defining the levels of abstraction to include in a model without presupposing the standard set.

Descriptions of Levels of Abstraction

This section provides detailed descriptions of Rasmussen's (1986) and Rasmussen et al.'s (1994) standard set of five levels of abstraction. The labels assigned to these levels are functional purposes, value and priority measures, purpose-related functions, object-related processes, and physical objects. These labels, which are those most recently suggested by Rasmussen, are documented in Reising (2000). Both these labels and the following descriptions of the concepts underlying each level, like those in Rasmussen et al.'s text, may be better suited to intentional rather than causal systems. Consequently, this account of the abstraction dimension complements Rasmussen's (1986) and Vicente's (1999) accounts, which are arguably better suited to causal systems.

Purposes and Constraints	Objectives: Cultural Mediation; Public Education; Assembly Place for Cultural Activities, Public Information Center Constraints: Budget Limits; No Political, Moral Censorship; Only High Quality Information; Union Agreement; General Work Regulations
Priority Measures, Abstract Functions	Flow of funds: sources and applications: Customer payment, fiscal law; Book purchase, Salaries Quality measures: literary and cultural values; Use of products: volume with reference to quality and population categories; Distribution across fact literature, fiction, and other materials
General Functions	User service; Book and art-item selection and purchase; Exhibition programs; Information Storage, Retrieval and Mediation; Administration: Books, employees; finance
Physical Processes and Activities	Reading books; talking to customers, finding, getting, storing books, use of computer tools and card files; etc.
Physical Resources and their Configuration	Librarians, reference librarians, cataloguers, clerks. Scientific and fiction book stock, card catalogs, handbooks, lexicographic tools, dictionaries, local and remote databases, Rooms for lending, reading and exhibitions. Shelving and storage facilities. Meeting and reference rooms. Administrative tools and facilities

FIGURE 3.11

Abstraction hierarchy of a library. (Adapted from Rasmussen, J., Pejtersen, A. M., and Goodstein, L. P., 1994, *Cognitive systems engineering*, Wiley, NY. Copyright © 1994 John Wiley & Sons. This material is reproduced with permission of John Wiley & Sons, Inc.)

That said, analysts have used variations of Rasmussen's most recent labels (Reising, 2000) in models of causal systems (e.g., Reising and Sanderson, 2002b[*]) and labels from Rasmussen (1986) in models of intentional systems (e.g., Hajdukiewicz et al., 1999[†]). Ultimately, the various labels are useful for understanding the basic concepts, although perhaps none of the labels are perfect. Devising a set of labels that is ideal for every analyst, system, or application is probably impossible. It is more important to concentrate efforts on learning the concepts that distinguish each level.

The following descriptions of the standard levels are illustrated with examples from four work domain models. Specifically, these are models of a home (Figure 2.7), a library (Rasmussen et al., 1994; Figure 3.11), a military command-and-control system for suppression of enemy air defense (SEAD) missions (Rasmussen, 1998; Figure 3.12), and a pasteuri-

[*] See Figure 5.3 for Reising and Sanderson's model of a causal system with Rasmussen's most recent labels.
[†] See Figures 5.6 and 6.8 for Hajdukiewicz et al.'s models of intentional systems with labels from Rasmussen (1986).

Goals and Purposes	Mission objectives within allocated resources: "immediate objective is to permit effective friendly air operations by protecting friendly airborne systems, disrupting cohesion of enemy air defenses, while respecting international conventions and protecting Air Force personnel and civilian population"
Priority Measures	Planning criteria: priority of combat versus SEAD Cost-effectiveness of mission: probability of success/loss/fratricide. (The enemy air defense order of battle, its system capabilities, and the flight profiles and defensive capabilities of projected friendly aircraft is used by the JFACC to develop a recommended threat priority list)
General Functions	Planning: conduct SEAD planning as directed by the JFC; develop intelligence requirements; support component commanders in developing planning priorities; allocate assets to conduct SEAD operations; request SEAD support from the JFC or other component commander; direct and control operations, monitor SEAD activities, Active operations: attack, destruction, disruption; Threat detection and identification; Coordination with surface support (e.g., field artillery, naval surface fire, surface-to-surface missiles)
Physical Processes	Functional characteristics of vehicles, F15, F16, F4-G, UCAV, URAV: - Speed & maneuverability (potential for evasive flight profiles, G limits & turning radii); - Vulnerability characteristics, thickness of armor, radiation characteristics, radar, IR. Functional characteristics of weapons: - destructive (Bombs, missiles, mines, artillery) and - disruptive (electromagnetic jamming and electromagnetic deception, expendables i.e., chaff, flares, and decoys). Functional characteristics of sensors: - Intelligence Collection (AN/APG-70, Lantirn, Pave Tack, PDF (ELINT)) - Threat detection and identification (AN/APG-70, Lantirn, ESM)
Inventory Configuration Topography	Map of theater territory with location and: Vehicle types, equipment, and numbers; Weapon types; Sensor types

FIGURE 3.12

Part of a work domain model of a military command-and-control system for SEAD missions. The full model is shown in Figure 4.10. (Adapted from Rasmussen, J., 1998, *Ecological interface design for complex systems: An example: SEAD–UAV systems* (AFRL-HE-WP-TR-1999-0011), Air Force Research Laboratory, Human Effectiveness Directorate, Wright-Patterson Air Force Base, OH. With permission.)

zation plant (Figure 3.13). The representation of the pasteurization plant was developed by Rasmussen but appears in Reising (1999).

Also noted in the discussions of the standard levels are some of the most common variations that analysts have employed in modeling the concepts at each level. These variations do not signify fundamental departures from the standard set but rather seem to reflect presentation choices. Differences in how the concepts in a model are presented may be due to a system's characteristics or a model's intended application.

Functional Purposes

The functional purposes level of abstraction represents constraints in terms of the *objectives* of a system as well as the *external limits* on the system's operation. Many systems have a very complex set of objectives, which may be viewed as comprising primary and secondary aims. A system's *primary objectives* reflect the fundamental reasons for which it exists, or is necessary, or its principal services or outputs. The main reason for which a home (Figure 2.7) exists is to promote *well-being,* whereas a library's (Figure 3.11) principal services include *cultural mediation* and *public education.* Similarly, the basic reason for which a SEAD military system (Figure 3.12) is necessary is *to permit effective friendly air operations by protecting friendly airborne systems and disrupting the cohesion of enemy air defenses,* while the principal service of a pasteurization plant (Figure 3.13) is to *pasteurize milk.*

A system's primary objectives usually leave many possibilities for action open to actors, which are reduced on the basis of its secondary objectives. The *secondary objectives,* which are also included at the functional purposes level, do not signify a system's principal services or outputs or the fundamental reasons for which it exists or is necessary, but still reflect its values in the form of what is to be attained or accomplished. The secondary objectives of a home include *environmental protection*; those of a SEAD military system include *protecting Air Force personnel and the civilian population*; and those of a pasteurization plant include *maximizing production, minimizing product waste and energy loss,* and *avoiding downtime.* Thus, the home, SEAD military system, and pasteurization plant must achieve their primary objectives while fulfilling these secondary aims, which remove options for action. Sometimes, a system's values may be formalized in organizational laws or regulations, or they may signify social conventions or norms.

As well as a system's primary and secondary objectives, the functional purposes level represents the *external constraints* on the system's operation. External constraints arise from the values of entities outside a system or from the values of society. Sometimes, these values, too, may be formalized in organizational laws or regulations. *Residential laws and regulations, union agreements,* and *international conventions* are examples of such external constraints on a home, library, and SEAD military system, respectively. At other times, the external constraints may be indicative of social conventions or norms. In the home model, *environmental protection* signifies the values not only of inhabitants but also of a town council, which provides recycled waste removal services to homes. Although there are no laws or regulations enforcing inhabitants to recycle their waste products, social conventions or norms place constraints on inhabitants to utilize the recycling services.

As the preceding discussion highlights, the functional purposes level may encompass a system's values as well as those of external entities or society. Often, there is commonality between the two sets of values. In the model of a home, *environmental protection* indicates the values of inhabitants as well as those of a town council. Moreover, the two sets of values may be shared across many different systems. For example, the values of *environmental*

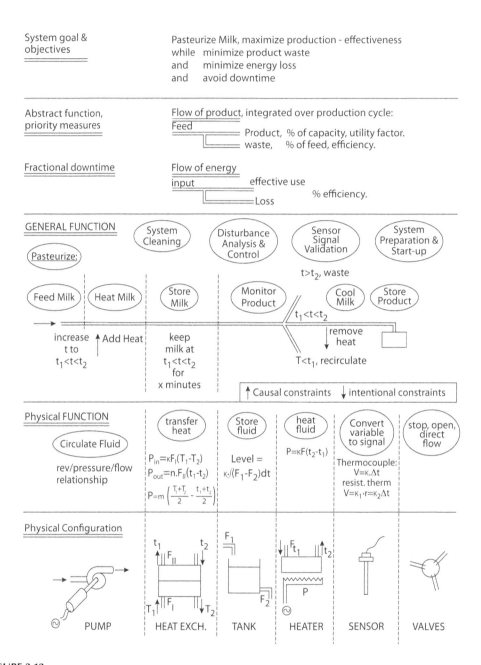

FIGURE 3.13

Abstraction hierarchy of a pasteurization plant developed by Rasmussen. (This figure is derived from a copy of a hand-drawn version by Rasmussen. Any mistakes are my own and are not attributable to Rasmussen. From Reising, D. V. C., 1999, *The impact of instrumentation location and reliability on the performance of operators using an ecological interface for process control*, unpublished doctoral dissertation, University of Illinois at Urbana-Champaign. With permission of the author.)

protection, *maximizing production*, and fair conditions for workers as specified in *union agreements* may be present in a variety of systems.

A system's primary objectives, secondary objectives, and external constraints are often conflicting. In the case of a home, *residential laws and regulations*, which prohibit inhabitants from adding a second floor to their home, may be incompatible with *well-being*. Likewise, with respect to a SEAD military system, *disrupting the cohesion of enemy air defenses* may conflict with *protecting Air Force personnel and the civilian population*. Last, in the context of a library, respecting *budget limits* may be at variance with providing *cultural mediation* and *public education*.

A system must be designed or assembled such that it is capable of fulfilling its primary and secondary objectives as well as satisfying its external constraints. For this reason, a system's functional purposes determine the physical objects that must be available in the system, the functionality of those physical objects, the functions the system must support, and the criteria the system must respect for effective performance. A system's functional purposes, therefore, provide the fundamental explanation for its design.

Rasmussen et al. (1994) note that the constraints at the functional purposes level "are in terms referring to properties of [the] environment" (p. 38). This means that these constraints emphasize the relationships between a system and its environment. Specifically, they highlight the objectives the system serves in the environment and the limits the environment places on the system. For instance, a home provides *well-being* for inhabitants, and the town council imposes *residential laws and regulations* on the home. In the same way, a library provides *cultural mediation* and *public education* for the community, and workers' unions negotiate *union agreements* with the library. Hence, the constraints at the functional purposes level govern the interaction between a system and its environment.

The most common variations in the way a system's functional purposes are modeled are that secondary objectives or external constraints are not represented at this level. In some models, these properties are expressed implicitly at the value and priority measures level in the form of criteria for effective performance. For instance, in the model of a home, *environmental protection* could have been expressed solely as a criterion such as *conservation of natural resources* at the value and priority measures level rather than also as a secondary objective or external constraint at the functional purposes level. In other models, the absence of these properties may reflect the characteristics of a system or the requirements of an application.

Finally, it is important to distinguish between a system's functional purposes and its goals or tasks. Goals or tasks vary as a function of the situation and, therefore, are dynamic (Burns and Vicente, 2001). For example, the goals or tasks of the inhabitants of a home depend on whether they are entertaining friends or cleaning the bathroom. Functional purposes, in contrast, remain relatively constant across a range of situations and thus are more stable over time. In the model of a home, the functional purpose of *well-being* is relevant regardless of whether inhabitants are entertaining friends or cleaning the bathroom. Given that the aim of work domain analysis is to produce an event-independent representation of a system, goals or tasks are not represented in an abstraction–decomposition space. Instead, they are the focus of control task analysis, the second dimension of cognitive work analysis, which models the constraints associated with recurring classes of situations. It is useful to bear in mind, however, that a system's functional purposes may be ascertained by examining what objectives or external constraints are relevant across a variety of tasks (see Burns and Hajdukiewicz, 2004, for some examples).

Value and Priority Measures

The value and priority measures level of abstraction represents the criteria that must be respected for a system to achieve its functional purposes. Criteria are fundamental *laws*, *principles*, or *values*, which can serve as a basis for evaluation or judgment. The criteria that must be respected for a home (Figure 2.7) to achieve *well-being* include *total income − total expenses = savings* and *pleasure*. Similarly, to realize their individual functional purposes, a library (Figure 3.11), a SEAD military system (Figure 3.12), and a pasteurization plant (Figure 3.13) must respect the criteria of *literary and cultural values*, *cost-effectiveness of a mission*, and *efficiency in the flow of energy*, respectively. Actors can employ the criteria at this level to evaluate how well the purpose-related functions of a system are fulfilling its functional purposes.

The criteria at this level also allow actors to compare, prioritize, and allocate resources to the purpose-related functions so that the system attains its functional purposes. For instance, in the case of the home model, the criteria of *total income − total expenses = savings* and *time* allow actors to consider how resources are best distributed across the purpose-related functions of *housework* and *maintenance*. Inhabitants may assess that, to achieve *well-being*, they can allocate money to *housework* and hire a cleaner, which would allow them to balance their *time* more effectively across work, rest, and leisure, but that they must perform *maintenance* themselves in order to respect the principle of *total income − total expenses = savings*. Likewise, in relation to the model of a library, the criteria of *flow of funds* and *literary and cultural values* enable actors to compare, prioritize, and allocate resources across the purpose-related functions of *book and art-item selection and purchase* and *exhibition programs*. For example, librarians may utilize these criteria to assess the number as well as the nature of the books and exhibition programs in which to invest money so that they can provide *cultural mediation* while still respecting their *budget limits*, and thus accomplish their functional purposes. The criteria at the value and priority measures level, then, may be viewed as shaping the flow of resources through a system.

To compare the effects of multiple purpose-related functions on a common set of functional purposes, the criteria at the value and priority measures level must be relatively abstract. As a result, these criteria are relevant not only to a range of purpose-related functions but also to many different systems. Time and money, for instance, are abstract concepts that are valued by a variety of systems. It is for this reason that Rasmussen et al. (1994) state that the representations at this level are "in abstract terms, referring neither to system nor environment" (p. 38).

Because the criteria at the value and priority measures level represent fundamental laws, principles, or values, the concepts at this level are relatively stable. For example, concepts such as *time*, *flow of funds*, and the prevention of the loss of human lives through *fratricide* in the models of a home, library, and SEAD military system, respectively, reflect fairly enduring human values. In this sense, the concepts at this level may be viewed as following "some kind of conservation law" (Rasmussen et al., 1994, p. 39). These concepts are not expected to "vanish" from a system either because physical or natural laws, such as those relating to mass or energy, cannot be abolished or because social laws, conventions, or values, such as those relating to people or money, are reasonably persistent (Rasmussen et al., 1994, p. 39).

The criteria at the value and priority measures level allow actors to reason from first principles. This type of reasoning is important, especially for dealing with novel or unanticipated situations. In the case of a home, inhabitants may routinely employ various heuristics to ensure that they respect the principle of *total income − total expenses = savings*, such as that grocery bills must be approximately $100 per week or recreation expenses must not exceed $80 per week. However, if inhabitants are confronted unexpectedly with substantial medical expenses, they can utilize the principle of *total income – total expenses = savings* to reason about what reductions are required to their grocery, recreation, or other expenses so that they can continue to fulfill their functional purposes. Likewise, librarians may employ certain heuristics, such as the names of well-known authors, to guide book purchases that are compatible with their *literary and cultural values*. When books are written by unknown authors, however, librarians can determine whether to purchase those books by assessing their contents against the library's *literary and cultural values*.

From the preceding discussion, it is evident that a system's values may be apparent at both the functional purposes and the value and priority measures levels. The main distinction between the two cases is that, at the functional purposes level, values are manifested in the form of what is to be attained or accomplished, whereas at the value and priority measures level, values are manifested in the form of criteria that enable the evaluation of whether a system is fulfilling its functional purposes. In the home model, values are apparent as secondary objectives or external constraints such as *environmental protection* as well as criteria for evaluation such as *conservation of natural resources*. Similarly, in the model of a SEAD military system, values are evident as secondary objectives such as *protecting Air Force personnel* and as criteria for evaluation such as *probability of fratricide*.

The most common variations in the way a system's value and priority measures are modeled relate to the nature of the representations at this level. Some models signify the criteria that must be respected for a system to achieve its functional purposes in terms of fundamental laws or principles. In the model of a home, *total income – total expenses = savings* is an example of this case. Other models describe criteria in the form of the values that must be preserved by a system. An example that can be directly compared with the preceding one is 'wealth.' Alternatively, some models depict criteria in terms of specific measures. One such description is 'expenses/income < 1.0.' Still other models denote criteria in the form of flows, balances, or distributions such as those relating to mass, energy, money, people, or information. These flows, balances, or distributions may be viewed as those that are necessary in a system to abide by particular laws, principles, or values. An example of this case is a representation of the 'flow of funds' that is needed in a system. Finally, some models describe criteria in terms of various general concepts, which may have specific laws, principles, values, measures, flows, balances, or distributions implicit within them. Examples of such descriptions are 'money,' 'funds,' or 'salaries.' Different types of representations are sometimes present within a single model.

Purpose-Related Functions

The purpose-related functions level of abstraction represents the functions that a system must be capable of supporting so that it can fulfill its functional purposes. *Preparation of meals and beverages*, *recreation*, *rest*, and *housework* are some of the functions that a home (Figure 2.7) must enable to achieve *well-being*. A library (Figure 3.11) must facilitate *book and art-item selection and purchase* and *information storage, retrieval and mediation* to realize *public education*. Similarly, a SEAD military system (Figure 3.12) must support *planning*, *active operations*, and *threat detection and identification* to *permit effective friendly air operations by protecting friendly airborne systems* and *disrupting the cohesion of enemy air defenses*. Last, a pasteurization plant (Figure 3.13) must afford *system cleaning, disturbance analysis and control*, and *sensor signal validation* in order to *pasteurize milk*.

The purpose-related functions level may be viewed as describing the "uses" that physical objects and their object-related processes are "put to" in a system (Miller and Vicente, 1998, p. 15). *Preparation of meals and beverages* points to the uses that a *refrigerator* and its *cooling capacity*, *utensils* and their *food and beverage handling capacity*, and a *stove* and its *heating capacity* serve in a home. In much the same way, *threat detection and identification* explains the uses of *sensors* and their *functional characteristics* in a SEAD military system, while *feed milk* indicates the uses of a *pump* and its capacity to *circulate fluid* in a pasteurization plant.

Typically, a variety of purpose-related functions are necessary to fulfill a system's functional purposes, and each function places demands on the system's resources. This means that the purpose-related functions must be coordinated in such a way that the functional purposes are achieved with the available resources. In a home, purpose-related functions such as *recreation*, *rest*, and *exercise* as well as *housework*, *maintenance*, and *professional work* must be managed in a fashion that attains *well-being* within the bounds of the system's resources. In a library, the purpose-related functions of *book and art-item selection and purchase*, *exhibition programs*, *information storage*, *retrieval and mediation*, and *administration* must be organized so that *cultural mediation* and *public education* are realized with the resources that are present. It is for this reason that Rasmussen et al. (1994) state that this level of abstraction represents the "Properties necessary and sufficient to identify the 'functions' which are to be coordinated" (p. 38).

A system's purpose-related functions are described in terms that are independent of the underlying physical objects or object-related processes by which they may be implemented. In a home, *recreation* may be realized by watching television, playing a board game, or reading a book, whereas *housework* may be accomplished by sweeping with a broom, dusting with a cloth, or vacuuming with a vacuum cleaner. In a library, *book and art-item selection and purchase* may be achieved by ordering items through a Web site or a mail order catalog, while in a SEAD military system, *destruction* may be attained with a variety of weapons. That is why Rasmussen et al. (1994) state that the concepts at this level are represented in "general" terms or in the form of "input–output relations" or "black box" models (pp. 38–39).

As a rule, each purpose-related function may be achieved with multiple physical objects and object-related processes, and the same physical resources may contribute to a variety of purpose-related functions. Therefore, if a physical resource is employed for a particular function, it may not be available for another one. The use of an armchair for watching television or *recreation*, for example, may mean that it is not available for sleeping or *rest*. On the other hand, if a physical resource is unavailable for a particular function, it may be possible to utilize another physical resource to achieve that function. For instance, instead of an armchair, a sofa may be used for sleeping or *rest*. Thus, the use of a system's physical objects and object-related processes must be orchestrated in a manner that fulfills the purpose-related functions.

Finally, the representations at the purpose-related functions level mirror the "familiar" language of the professional field(s) relevant to a system (Rasmussen et al., 1994, p. 38). *Destruction*, *disruption*, and *threat detection and identification* are common terms in the military field that are applicable to a range of military systems. In the process control industry, terms such as *system cleaning*, *disturbance analysis and control*, *sensor signal validation*, and *system preparation and start-up*, as used in the pasteurization plant model, are pertinent to many systems. Likewise, *preparation of meals and beverages*, *dining*, *housework*, and *maintenance* are suitable terms for describing a variety of homes. Hence, the representations at this level indicate the type of system but do not reveal the specific system, such as a particular military system, pasteurization plant, or home.

Object-Related Processes

The object-related processes level of abstraction represents the functional processes or the functional capabilities or limitations of a system's physical objects. In a home (Figure 2.7), the functional capabilities or limitations of physical objects include the *washing capacity* (e.g., load limits, washing cycles, and temperature range) of a dishwasher, the *showering capacity* of a shower, and the *data handling capacity* of a computer. Some of the functional processes of a library (Figure 3.11) are *reading books* and *talking to customers*, while the object-related processes of a SEAD military system (Figure 3.12) include the *functional characteristics of vehicles*, such as their *speed and maneuverability*. Examples of the functional processes of a pasteurization plant (Figure 3.13) are *storing* and *heating fluid*.

A system's object-related processes serve to achieve its purpose-related functions. In a home, the *showering capacity* of a shower enables the purpose-related function of *personal care and grooming*, while the *data handling capacity* of a computer permits *professional work*. The *functional characteristics of vehicles* in a SEAD military system allow *active operations* to be accomplished, and in a pasteurization plant, the capacity for *storing* and *heating fluid* makes the purpose-related functions of *storing* and *heating milk* possible.

The potential uses of a system's object-related processes for achieving its purpose-related functions are often implicit in the representations at the object-related processes level. In the home model, the descriptions *serving and seating capacity* and *sleeping capacity* suggest that these object-related processes may be used for *dining* and *rest*, respectively. Similarly, in a pasteurization plant, the descriptions *storing fluid* and *heating fluid* highlight the uses of these functional processes for *storing* and *heating milk*, respectively.

The object-related processes level encompasses the functional processes or the functional capabilities or limitations of both natural and artificial physical objects. For instance, the model of a library includes the functional process of *talking to customers*, which is afforded by librarians. In addition it includes the functional process of *finding books*, which is made possible not only by librarians but also by artificial objects such as card catalogs.

Object-related processes are highly dependent on the properties of physical objects, or causal properties, and cannot be altered unless the properties of those objects are changed. For example, the *serving and seating capacity* of a table in a home is highly dependent on the size of the table and cannot be varied without modifying the table's properties. This is distinguished from purpose-related functions, which are independent of the underlying physical objects by which they may be implemented. For instance, the purpose-related function of *dining* in a home may be achieved while seated at a table, on a bed, or in an armchair. Compared with object-related processes, therefore, purpose-related functions reflect greater intentionality in how a system's physical objects are utilized.

Another distinction between object-related processes and purpose-related functions is that the former highlight what a system's physical objects can do, or can afford, whereas the latter emphasize the uses that these objects are commonly put to in a system. In the model of a home, the object-related process of *seating capacity* suggests that an armchair can afford sitting in, jumping on, jumping off, or standing on. On the other hand, the purpose-related functions of *dining*, *rest*, and *recreation* indicate the regular uses of an armchair in a home. The purpose-related functions reduce the possibilities for action afforded by physical objects and their object-related processes in a way that is necessary for achieving the functional purposes.

The most significant variation in the way a system's object-related processes are represented is that some models list the names of physical objects at this level. Burns and

Hajdukiewicz (2004) note that the names of physical objects in this context serve as tags or placeholders for more detailed statements of their functional processes or functional capabilities or limitations. In addition, Miller and Vicente (1998) suggest that the names of physical objects in this context reflect their roles in a system and thereby how these objects serve to achieve the system's purpose-related functions. In such models, the bottommost level of abstraction is reserved for representing the physical attributes (e.g., appearance, material properties, location, spatial distribution) of the objects named at the higher level.

Physical Objects

The physical objects level of abstraction represents the physical objects of a system. A home's (Figure 2.7) physical objects encompass such items as a *dishwasher, refrigerator, stove, television,* and *telephone*. The physical objects of a library (Figure 3.11) include *librarians, scientific and fiction book stock, local and remote databases,* and *rooms for lending, reading, and exhibitions*. In a SEAD military system (Figure 3.12), the physical objects include *vehicles, weapons,* and *sensors*, while some of the physical objects in a pasteurization plant (Figure 3.13) are a *pump, tank, heat exchanger,* and *heater*.

A system's physical objects afford functional processes or functional capabilities, which serve to achieve its purpose-related functions. In a home, a *refrigerator* and a *stove* provide functional capabilities relating to *cooling* and *heating*, respectively. Likewise, in a library, the *scientific and fiction book stock* affords the functional process of *reading books*, whereas the *local and remote databases* afford the functional process of *finding books*. Last, in a pasteurization plant, a *pump* and a *heat exchanger* allow the functional processes of *circulating fluid* and *transferring heat*, respectively, to be realized.

The physical objects level may represent artificial objects such as equipment, tools, artifacts, or infrastructure. The model of a home lists equipment, including a *refrigerator* and a *computer,* and the model of a library contains infrastructure such as *rooms for lending, reading, and exhibitions*. As well as artificial objects, this level may depict geographical features, including vegetation and land elevation, and other natural entities such as people. The model of a library, for instance, lists *librarians* and *clerks*.

The representation of physical objects at this level may include the following information about each object:

1. Name
2. Number or amount
3. Type (e.g., model number, style)
4. Appearance (e.g., shape, color, dimensions)
5. Configuration (i.e., arrangement of the object's parts or elements)
6. Material properties (i.e., substances of which the object is made or composed)
7. Location or position
8. Topography or spatial distribution
9. Physical connections

These and other physical attributes may be depicted at this level in many forms, including text; photographs; drawings (e.g., architectural drawings of buildings, mechanical drawings of equipment); diagrams (e.g., circuit diagrams of electrical processes); maps (e.g., work site maps, land elevation maps); and videos. However, the majority of models simply list the names of physical objects at this level unless more detail is required for an application.

The physical objects level may be viewed as representing the physical world or physical reality of a system. This information is necessary for activities such as navigating in a physical space and finding, distinguishing, and identifying objects. That is why Rasmussen et al. (1994) state that this level represents the "Properties necessary and sufficient for classification, identification and recognition of particular material objects and their configuration: for navigation in the system" (p. 38).

Summary

An important decision in building a work domain model of a system is the number, types, and labels of the levels of abstraction to include in the model. This chapter reviewed representations of numerous systems to establish whether analysts can adopt Rasmussen's (1986) and Rasmussen et al.'s (1994) set of five levels of abstraction to model any system, or whether it may be wise for analysts not to presume this conventional set a priori. The review highlighted that this set has been both useful for modeling a range of systems and beneficial for a variety of applications. Therefore, detailed descriptions were provided of this set, illustrated with examples from models of a home, a library, a military system, and a pasteurization plant. The guidelines for work domain analysis in Chapter 8 also provide a number of prompts and keywords for analyzing the constraints to include in a model based on the standard levels.

However, the review also highlighted that one reason for the strong similarity between the levels of abstraction in a range of models and the standard set may be that analysts were working from example. Although these levels have been useful in numerous contexts, they may not be the most suitable set for all systems. In such cases, an alternative set that offers a more valid or faithful representation of a system is preferable as it is likely to lead to better designs or products for that system. Models of the games of chess and baseball (Vicente and Wang, 1998), for example, comprise at least two levels that appear to be different from the standard levels. That is why the guidelines for work domain analysis in Chapter 8 also suggest a strategy for developing a model of a system without presupposing the standard set.

4

Decomposition

OVERVIEW   Compared with the abstraction dimension, the decomposition dimension of a work domain model is underrated with respect to its significance. Yet, as this chapter shows, this dimension is essential for efficiently constructing a well-formed model that comprises the most meaningful or useful representations of a system. The chapter begins by examining existing models of a variety of systems to establish whether there are patterns in the number, types, and labels of the levels of decomposition that analysts can exploit for building new models. Following that, two approaches are described for creating representations of a system at different levels of detail, namely, constraint decomposition and system decomposition. Finally, the chapter explains why defining the decomposition dimension of a work domain model methodically—using system decomposition—is important, even when an abstraction hierarchy appears sufficient for an application.

Number, Types, and Labels of Levels of Decomposition

In developing a work domain model of a system, it is useful for analysts to consider the number, types, and labels of the levels of decomposition to include in the model. As for the abstraction dimension, the two main concerns of analysts in making these decisions are to create a valid or faithful representation of the system and to produce a model that is fit for its intended application. Work domain models that disregard these considerations may lead to designs or products that do not cultivate system safety, productivity, or workers' health.

As highlighted in Chapter 3, since Rasmussen (1979, 1985) formulated the basic concepts of work domain analysis, analysts have produced work domain models of many kinds of systems (see Table 3.1). Of these models, few are in the form of the abstraction–decomposition space compared with the abstraction hierarchy. Despite this, there are perhaps just enough cases of the former to make it worth examining whether there are any emerging patterns in the number, types, and labels of the levels of decomposition in these models. If so, analysts will be able to exploit these patterns in building new models. Otherwise, analysts will need to proceed without any preconceptions about the nature of the decomposition dimension that should be included in their models.

It is important to state from the outset that, unlike the situation for the abstraction dimension, there is no standard set of levels of decomposition for modeling a variety of systems. Instead, as this section shows, there is significant variation in the number, types, and labels of the levels of decomposition in models across a range of systems, although one pattern that is evident relates to whether this dimension has a physical or a conceptual basis. The variations are most likely due to the characteristics of a system or the requirements of an application.

Briefly, the number of levels of decomposition in existing models varies from as few as two to as many as seven. Even in models of systems belonging to the same class, the number of levels usually varies. Consider, for example, four models of the human body. One model by Sharp (1996), which was discussed previously,[*] has four levels: body, body systems, body organs, and cellular. Another model by Hajdukiewicz (1998), also presented previously,[†] comprises five levels: whole body, system, organ, tissue, and cell. Watson and Sanderson's (1998) model has four levels that are identified as system, subsystem, component, and part. Last, Thompson et al.'s (2003) model consists of two levels that are named whole body and organ. These four models were intended for markedly different applications, which would have shaped how the analysts characterized the human body.

Similarly, the types of levels of decomposition in models of various systems differ widely. This is because they tend to be based on the functional parts of specific systems. For instance, the levels may be defined by the functional parts of the human body (Hajdukiewicz, 1998; Sharp, 1996; Thompson et al., 2003); a thermalhydraulic microworld simulation[‡] (Bisantz and Vicente, 1994); the fuel and engine systems of an aircraft (Dinadis and Vicente, 1999); or an emergency management organization[§] (Rasmussen et al., 1987).

The labels for the levels of decomposition are also usually based on specific systems. In most models of the human body, the labels stem from the names of body parts (Hajdukiewicz, 1998; Sharp, 1996; Thompson et al., 2003). Likewise, in the model of a home (Figure 2.7), the labels (i.e., whole house, rooms or subspaces, and contents) are derived from common names for the functional parts of this system.

One underlying pattern in the levels of decomposition across a variety of systems, however, is that they have either a physical or a conceptual basis. The majority of models have a decomposition dimension with a *physical basis*, which means that the levels are based on functional aggregations of physical objects. A few models have a decomposition dimension with a *conceptual basis*, which means that the levels are defined by functional aggregations of a mixture of concepts, perhaps including physical objects.

One model characterized by physical decomposition was produced by Jamieson and Vicente (2001) for a process control system, specifically a simulation of a fluid catalytic cracking unit in a petrochemical refinery (also see Jamieson, 1998). Figure 4.1a, which provides an overview of the abstraction–decomposition space they developed, shows that the three levels of decomposition in this model are system, unit, and component. Figure 4.1b specifies the functional parts at each level. The system level represents the entire fluid catalytic cracking unit. At the unit level, the functional parts include the lift air supply unit, the regenerator unit, and the catalyst circulation unit. The functional parts at the component level, which are depicted diagrammatically, include blowers, valves, and heaters. Hence, the levels of decomposition in this model are defined by functional aggregations of physical objects.

Other models that are defined by physical decomposition include representations of the human body[¶] (Hajdukiewicz, 1998; Sharp, 1996; Thompson et al., 2003) and aviation systems (Dinadis and Vicente, 1999; Nadimian, Griffiths, and Burns, 2002; Xu, 2007). Most models of process control systems also fall in this category. Some examples are representations of a thermalhydraulic microworld simulation[**] (Bisantz and Vicente, 1994); a feedwater subsystem of a nuclear power plant (Dinadis and Vicente, 1996); a simulated coal-fired power plant (Burns, 2000); and an acetylene hydrogenation reactor in a petrochemical

---

[*] See Figure 3.7a for Sharp's model of the human body.
[†] See Figure 3.8a for Hajdukiewicz's model of the human body.
[‡] See Figure 2.15 for the model of the thermalhydraulic microworld simulation.
[§] See Figures 6.13 and 6.14 for models of the emergency management organization.
[¶] See Figures 3.7a and 3.8a for Sharp's and Hajdukiewicz's models of the human body, respectively.
[**] See Figure 2.15 for the model of the thermalhydraulic microworld simulation.

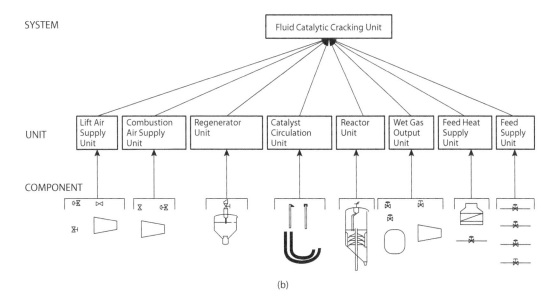

	System (S)	Unit (U)	Component (C)
Functional Purpose (FP)	FP-S		
Abstract Function (AF)		AF-U	AF-C (mass) AF-C (energy)
Generalized Function (GF)		GF-U	GF-C
Physical Function (PFn)			PFn-C
Physical Form (PFo)			

(a)

(b)

FIGURE 4.1

(a) Overview of an abstraction–decomposition space for a simulation of a fluid catalytic cracking unit in a petrochemical refinery. Representations were not developed at the physical form level of abstraction because the model is of a computer simulation. (Adapted from Jamieson, G. A., and Vicente, K. J., 2001, Ecological interface design for petrochemical applications: Supporting operator adaptation, continuous learning, and distributed, collaborative work, *Computers & Chemical Engineering*, 25, 1055–1074. Copyright © 2001. With permission from Elsevier.) (b) A specification of the parts at each level of decomposition in the model of the fluid catalytic cracking unit. (Adapted from Jamieson, G. A., 1998, *Ecological interface design for petrochemical processing applications,* unpublished master's thesis, University of Toronto, Ontario, Canada. With permission of the author.)

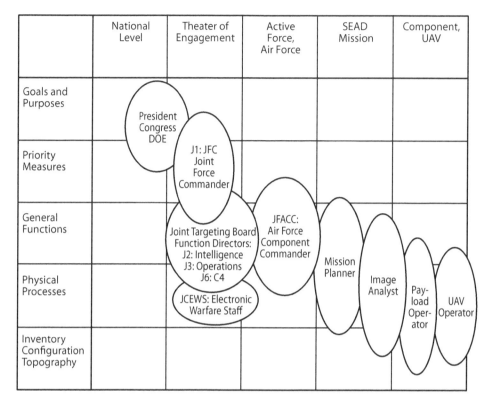

FIGURE 4.2

Overview of an abstraction–decomposition space for a military command-and-control system. The full model is shown in Figure 4.10. Part of this model was presented previously in Figure 3.12. (Adapted from Rasmussen, J., 1998, *Ecological interface design for complex systems: an example: SEAD–UAV systems* (AFRL-HE-WP-TR-1999-0011), Air Force Research Laboratory, Human Effectiveness Directorate, Wright-Patterson Air Force Base, OH. With permission.)

plant[*] (Miller and Vicente, 1998). The model of a home in Figure 2.7 is also characterized by physical decomposition.

One example of a model with a decomposition dimension that has a conceptual basis, on the other hand, was developed by Rasmussen (1998) for a military command-and-control system, specifically as it relates to the role of uninhabited aerial vehicles (UAVs) in suppression of enemy air defense (SEAD) missions. Figure 4.2 provides an overview of Rasmussen's model. This figure indicates that the five levels of decomposition in the model are national level, theater of engagement, active force (Air Force), mission (SEAD), and component (UAV). These levels do not reflect part–whole relationships as plainly as is the case when the decomposition dimension has a physical basis. Nevertheless, higher levels in the model provide coarse representations of the military command-and-control system, whereas lower levels provide fine-grained ones. The bubbles overlaid onto the model indicate those areas of the work domain with which particular organizational units or actors are concerned. Thus, the decomposition dimension reflects the "foci of attention" of the command structure or, in other words, the levels of resolution that the command structure adopts for viewing the work domain (Rasmussen, 1998, p. 52).

---

[*] See Figure 5.7 for the decomposition hierarchy of the acetylene hydrogenation reactor in a petrochemical plant.

Two other models that are defined by conceptual decomposition are representations of an emergency management system[*] (Rasmussen et al., 1987) and the French sea fishing industry (Morel and Chauvin, 2006).

In summary, analysts have not employed a standard set of levels of decomposition for modeling an array of systems. Instead, the number, types, and labels of the levels vary according to the type of system or the application. Although one emerging pattern is that the decomposition dimensions of existing models tend to have either a physical or a conceptual basis, there are insufficient cases to conclude definitively that these are the only types and that one of these types will be applicable to all systems. Therefore, while the guidelines for work domain analysis in Chapter 8 recognize this pattern, they provide a strategy for defining the decomposition dimension of a model without any preconceptions about its characteristics.

What to Decompose?

Irrespective of the number, types, and labels of the levels of decomposition in a work domain model, there are two approaches that analysts have utilized to formulate representations of a system at different levels of detail. One approach, which I label *constraint decomposition*, is to decompose the constraints in a model into more detailed representations of a work domain or to aggregate them into less detailed ones. The classic approach, which I call *system decomposition*, is to decompose a system into parts or aggregate the parts of a system into wholes to produce a decomposition hierarchy; full or partial abstraction hierarchies are then created for the parts at each level of decomposition. The rest of this section explains these approaches further. Then, in the final section of this chapter, the advantages of system decomposition over constraint decomposition are highlighted.

Constraint Decomposition

Constraint decomposition involves decomposing or aggregating the *constraints* in a model into respectively more or less detailed representations of a work domain. Typically, this approach is applied when a model is constructed in the form of the abstraction hierarchy. Figure 4.3 illustrates constraint decomposition generically; the nodes in this figure denote constraints. The shaded nodes are at a higher, or coarser, level of decomposition relative to the clear nodes nested within them. This type of model is crafted either by decomposing the shaded nodes to create more detailed representations of those constraints or by aggregating the clear nodes to develop less detailed representations of those constraints. Both decomposition and aggregation may be applied within a single model, and the constraints may be decomposed or aggregated multiple times, not just once as shown in Figure 4.3.

With this approach, the levels of decomposition of the constraints in a model are usually not defined systematically or explicitly. Therefore, the constraints within a particular level of abstraction may fall at different levels of decomposition, which, as explained later in this chapter, may not be ideal. For instance, in relation to Figure 4.3, it is possible that C_2 and

---

[*] See Figures 6.13 and 6.14 for models of the emergency management system.

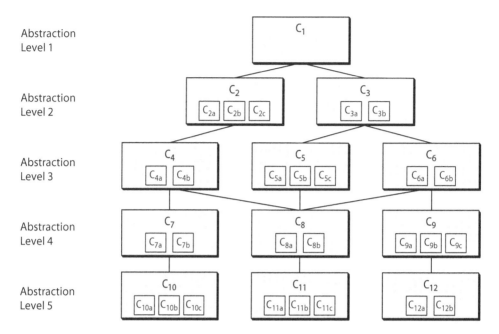

FIGURE 4.3
Constraint decomposition. C, constraint.

C_3 are at different levels of decomposition, despite being at the same level of abstraction. Similarly, C_{2a}, C_{2b}, and C_{2c} may not be at the same level of decomposition.

One model that adopts constraint decomposition was built by Lintern (2006) for insurgency operations (Figures 4.4 and 4.5). Figure 4.4 depicts the top three levels of abstraction in the model. At the third level (i.e., general mission functions), the shaded nodes portray information about insurgency operations at higher levels of decomposition relative to the clear nodes nested within them. For example, *tactical level of war* is shown decomposed into *violence, oppression & coercion* and *tactical enclaves*. Figure 4.5 depicts the bottom three levels of abstraction in the same model. The first of these (i.e., general mission functions) is the same as the lowest level of abstraction presented in Figure 4.4 but focuses on just one of the nodes, namely, *violence, oppression & coercion*. Specifically, this node is represented at lower levels of decomposition with items such as *leadership* and *plans*. The latter is decomposed further into four types of plans. In Figure 4.5, some of the nodes at the last two levels of abstraction (i.e., 'technical functions contextual effects' and 'physical resources and constraints') are also decomposed further than others. For example, at the physical resources and constraints level, *anti-insurgency forces* is not decomposed further and thus is shown at only one level of decomposition, whereas *area of operation* is shaded to indicate four levels of decomposition. Lintern notes that the decompositions in his model were taken to levels suggested by subject matter experts.

Other examples of models that adopt constraint decomposition include those of a simulation of a pasteurization plant[*] (Reising and Sanderson, 2002b); a fighter aircraft (Naikar and Sanderson, 1999); a multisensor command-and-control aircraft (Lintern and Miller, 2003); and a naval surface ship[†] (Linegang and Lintern, 2003).

---

[*] See Figure 5.3 for the model of the microworld simulation of a pasteurization plant.
[†] See Figure 6.6 for the model of the naval surface ship.

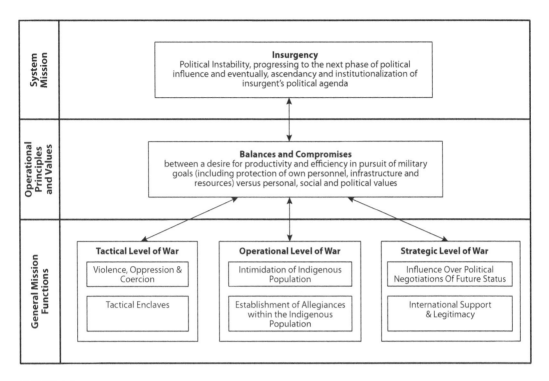

FIGURE 4.4

Top three levels of abstraction in a model of insurgency operations. (Adapted from Lintern, G., 2006, A functional workspace for military analysis of insurgent operations, *International Journal of Industrial Ergonomics*, *36*, 409–422. Copyright © 2006. With permission from Elsevier.)

System Decomposition

System decomposition involves decomposing a *system* into parts or aggregating the *parts* of a system into wholes so that a decomposition hierarchy can be formed. Subsequently, this hierarchy is used as a structure for formulating either more or less detailed representations of a work domain according to the parts at each level of decomposition. Specifically, progressively more detailed representations are created from higher to lower levels. Unlike constraint decomposition, therefore, this approach requires the decomposition dimension of a model to be specified precisely.

System decomposition has two variations, which are depicted generically in Figure 4.6. Both variations require the construction of a decomposition hierarchy to delineate the horizontal axis of an abstraction–decomposition space. Hence, a decomposition hierarchy is shown at the top of both models in Figure 4.6. The same hierarchy is presented in each case. This hierarchy has three levels of decomposition and one or more parts (e.g., 1A, 2A, 2B, 3C, 3D) at each level.

Once the decomposition dimension of a model has been specified, work domain representations are created at each level of decomposition using full or partial abstraction hierarchies. The first variation of system decomposition (Figure 4.6a) involves developing full or partial abstraction hierarchies for *all or many parts* of a system at each level of decomposition. This variation results in a comprehensive work domain representation encompassing several parts of a system at each level. The second variation (Figure 4.6b) involves developing full or partial abstraction hierarchies for only a *single part* of a system at each

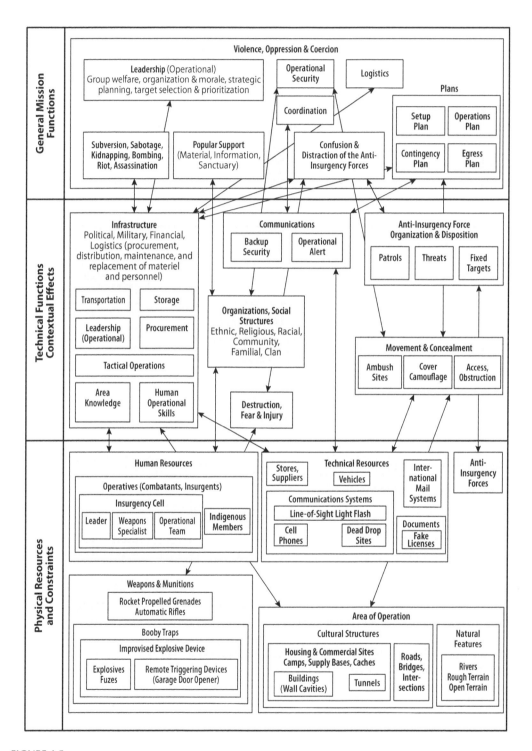

FIGURE 4.5

Bottom three levels of abstraction in a model of insurgency operations. (Adapted from Lintern, G., 2006, A functional workspace for military analysis of insurgent operations, *International Journal of Industrial Ergonomics*, *36*, 409–422. Copyright © 2006. With permission from Elsevier.)

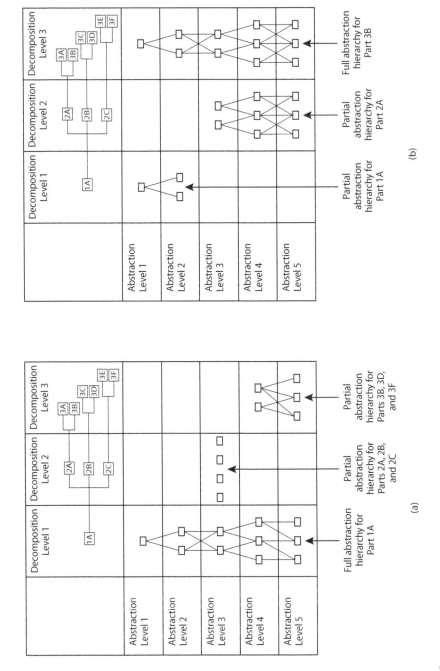

FIGURE 4.6

System decomposition, which has two variations: (a) first variation, in which full or partial abstraction hierarchies are developed for *all or many parts* of a system at each level of decomposition; (b) second variation, in which full or partial abstraction hierarchies are created for a *single part* of a system at each level of decomposition.

level of decomposition. This variation results in a detailed work domain representation of a particular part of a system at each level. Both variations may be applied within a single model, although at different levels of decomposition.

The first variation was adopted for the model of a home (Figure 2.7). To produce a decomposition hierarchy, the whole house was decomposed into some of its rooms or subspaces, and the rooms or subspaces were decomposed into some of their contents, as shown in Figure 4.7. Then, a partial abstraction hierarchy was composed for all or many parts of the home at each level of decomposition (Figure 2.7). At the whole house level, which comprises a single part, a partial representation (functional purposes and value and priority measures) takes into account the entire home as a single entity. At the rooms or subspaces level, a partial representation (purpose-related functions) encompasses many of the rooms or subspaces of the home. In the same way, at the contents level, a partial representation (object-related processes and physical objects) covers several of the contents of the home.

In the case of the home model, the three partial representations may be assembled to form a single abstraction hierarchy, spanning multiple levels of decomposition. This is straightforward because representations were produced for only one cell at each level of

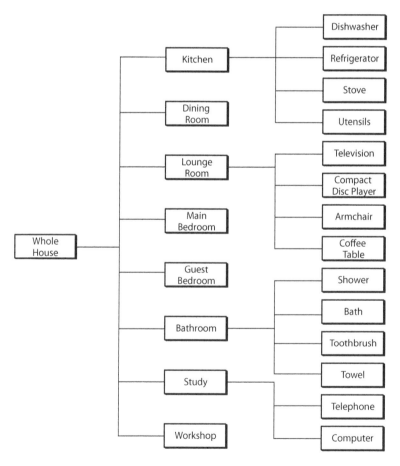

FIGURE 4.7
Decomposition hierarchy for the model of a home.

abstraction. Sometimes, however, analysts generate representations for more than one cell at each level of abstraction. Jamieson and Vicente's (2001) model in Figure 4.1a, for example, contains representations in two cells at some of the levels of abstraction (i.e., abstract function and generalized function). In this situation, the most appropriate cell from each level, given the application, may be selected for presentation in the form of the abstraction hierarchy.

Although the second variation of system decomposition was not employed for building the model of a home (Figure 2.7), this approach may be illustrated in the context of this model. That is, it is possible to discuss what representations would have been created at each level of decomposition in the model had the second variation been adopted. At the whole house level, a full or partial abstraction hierarchy would have been prepared for the entire home as a single entity, as was done for the first variation. At the rooms or subspaces level, a full or partial abstraction hierarchy would have been composed for only one of the rooms or subspaces of the home, for example, the kitchen. Last, at the contents level, a full or partial abstraction hierarchy would have been developed for just one of the contents of the home, for instance, the dishwasher.

One model based on the first variation of system decomposition is that constructed by Jamieson and Vicente (2001) for a simulation of a fluid catalytic cracking unit in a petrochemical refinery (also see Jamieson, 1998), which was introduced previously. To define the decomposition dimension of this model, the components of the fluid catalytic cracking unit were aggregated into units, and the units were aggregated into the whole system (Figures 4.1a and 4.1b). The shaded cells in Figure 4.1a indicate that the analysts built partial abstraction hierarchies for the parts at each level of decomposition. More specifically, partial representations were formulated for *all* of the parts at each level of decomposition. Figure 4.8 shows the representation for the cell at the generalized function level of abstraction and the unit level of decomposition (i.e., the cell marked GF-U in Figure 4.1a). This representation accounts for the generalized functions of all of the parts at the unit level of decomposition (Figure 4.1b). Figure 4.9 portrays the representation for the cell at the generalized function level of abstraction and the component level of decomposition (i.e., the cell marked GF-C in Figure 4.1a), which covers the generalized functions of all of

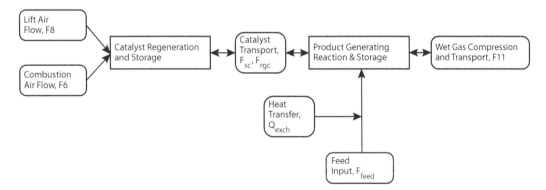

FIGURE 4.8

Representation for the cell at the generalized function level of abstraction and unit level of decomposition in a model of a simulation of a fluid catalytic cracking unit in a petrochemical refinery. (From Jamieson, G. A., and Vicente, K. J., 2001, Ecological interface design for petrochemical applications: Supporting operator adaptation, continuous learning, and distributed, collaborative work, *Computers & Chemical Engineering, 25*, 1055–1074. Copyright © 2001. With permission from Elsevier.)

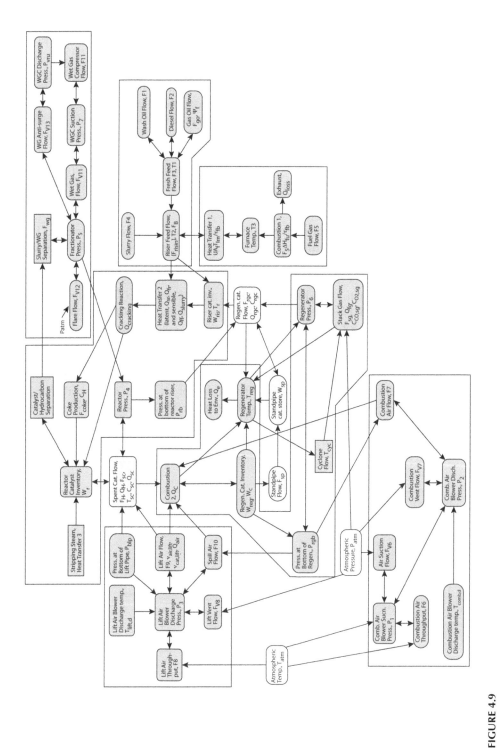

FIGURE 4.9

Representation for the cell at the generalized function level of abstraction and component level of decomposition in a model of a simulation of a fluid catalytic cracking unit in a petrochemical refinery. (From Jamieson, G. A., and Vicente, K. J., 2001, Ecological interface design for petrochemical applications: Supporting operator adaptation, continuous learning, and distributed, collaborative work, *Computers & Chemical Engineering*, 25, 1055–1074. Copyright © 2001. With permission from Elsevier.)

the parts at the component level of decomposition (Figure 4.1b). In both Figures 4.8 and 4.9, the arrows connecting the nodes depict causal relationships or causal flows between the generalized functions; these types of relationships are discussed in more detail in Chapter 5.

Some other models that adopt the first variation of system decomposition are those of a thermalhydraulic microworld simulation* (Bisantz and Vicente, 1994); the fuel and engine systems of an aircraft (Dinadis and Vicente, 1999); and an acetylene hydrogenation reactor in a petrochemical plant† (Miller and Vicente, 1998).

A model that adopts the second variation of system decomposition is that produced by Rasmussen (1998) for a military command-and-control system. This model, which relates to the role of UAVs in SEAD missions, was also presented previously (Figure 4.2). As indicated in Figure 4.10, the decomposition dimension of this model was specified by decomposing the military command-and-control system into national level, theater of engagement, active force (Air Force), mission (SEAD), and component (UAV). Then, full or partial abstraction hierarchies were created for only a *single* part at each level of decomposition; that is, a single national level (United States of America), a single theater of engagement (Balkan peacekeeping), a single force (Air Force), a single mission (SEAD), and a single component (UAV).

Another work domain model that adopts the second variation of system decomposition is that developed by Morel and Chauvin (2006) for the French sea fishing industry.

Why Decompose?

For many applications, the decomposition dimension of a work domain model may not appear to be essential. This may explain why significantly fewer models are in the form of the abstraction–decomposition space compared with the abstraction hierarchy. However, as this section shows, there are important reasons why analysts should not neglect a methodical analysis of the decomposition dimension, even when it seems unnecessary for an application. These reasons are that it will allow them to examine which representations of a work domain are the most meaningful or useful to create, given the purpose of the analysis, and it will help them to produce models with greater comprehensiveness, accuracy, and efficiency. As these motives presume system decomposition, this section highlights some of the benefits of this approach, over constraint decomposition, for formulating representations of a system at different levels of detail.

Producing an abstraction hierarchy of a system without defining the decomposition dimension of the model has a number of pitfalls. In Chapter 2, it was shown that the abstraction hierarchy differs from the abstraction–decomposition space mainly in that it does not indicate the levels of decomposition of the constraints in the model. Nevertheless, all of the constraints in an abstraction hierarchy do fall at some level of decomposition; this is just not depicted explicitly in the model.

---

* See Figure 2.15 for the model of the thermalhydraulic microworld simulation.
† See Figure 5.7 for the decomposition hierarchy of the acetylene hydrogenation reactor in a petrochemical plant.

	National Level	Theater of Engagement	Active Force, Air Force	SEAD Mission	Component, UAV
Goals and Purposes	Peace, Human rights, trade, Objectives of national policy and international treaties	Balkan peace keeping within stated policy and allocated resources, considering international relations and public opinion	Mission objectives within allocated resources for military engagement or humanitarian missions, respecting international conventions, while protecting military personnel and civilian population, and considering political and public opinion	Mission objectives within allocated resources: "Immediate objective is to permit effective friendly air operations by protecting friendly airborne systems, disrupting cohesion of enemy air defenses, while respecting international conventions and protecting Air Force personnel and civilian population"	Objectives: support of operations according to mission plans
Priority Measures	Trade deficit, level of threat to citizens, public opinion	Number of refugees and lost lives, threat to US citizens, public opinion, votes in Congress	Cost-effectiveness of missions: probability of success / loss / fratricide. (The enemy air defense order of battle, its system capabilities, and the flight profiles and defensive capabilities of projected friendly aircraft is used by the JFACC to develop a recommended threat priority list)	Planning criteria: priority of combat versus SEAD. Cost-effectiveness of mission: probability of success / loss / fratricide. (The enemy air defense order of battle, its system capabilities, and the flight profiles and defensive capabilities of projected friendly aircraft is used by the JFACC to develop a recommended threat priority list)	Cost-effectiveness of mission: probability of success / loss; Threat priority list from JFACC
General Functions	UN-intervention, diplomacy, trade-boycott, military intervention	Humanitarian help: Transportation of personnel, food, medicine, etc. Diplomacy, Military intervention: Protection of humanitarian services, monitoring activities of military and para-military activities, intervention in conflicts, intelligence	Organize commands and forces and employ those forces as necessary to accomplish assigned missions; develop objectives and guidance for the joint operation or campaign and specify the roles of air, land, maritime, space, and special operations forces in the conduct of the joint operation or campaign; establish requirements for SEAD to facilitate these operations; Surveillance, monitoring, intelligence (AWACS, JSTARS, UAV). Transport, supply, Combat (SEAD, etc.)	Planning: conduct SEAD planning as directed by the JFC; develop intelligence requirements; support component commanders in developing planning priorities; allocate assets to conduct SEAD operations; request SEAD support from the JFC or other component commander; direct and control operations, monitor SEAD activities, Active operations: attack, destruction, disruption; Threat detection and identification; Coordination with surface support (e.g., field artillery, naval surface fire, surface-to-surface missiles)	Planning: collect and distribute intelligence on enemy air defenses, nominate SEAD targets; monitor SEAD mission results; forward mission results to the JFC and other component commanders; Mission planning and control, control of UAV flights; Collection and distribution of battlefield intelligence and battle damage information; Air strike and combat guidance, area searches, route reconnaissance, target location

FIGURE 4.10

Abstraction–decomposition space of a military command-and-control system for SEAD missions. (Adapted from Rasmussen, J., 1998, *Ecological interface design for complex systems: An example: SEAD–UAV systems* (AFRL-HE-WP-TR-1999-0011), Air Force Research Laboratory, Human Effectiveness Directorate, Wright-Patterson Air Force Base, OH. With permission.)

Continued

	National Level	Theater of Engagement	Active Force, Air Force	SEAD Mission	Component, UAV
Physical Processes			Resources as specified at the lower decomposition levels	Functional characteristics of vehicles, F15, F16, F4-G, UCAV, URAV: - Speed & maneuverability (potential for evasive flight profiles, G limits & turning radii); - Vulnerability characteristics, thickness of armor, radiation characteristics, radar, IR. Functional characteristics of weapons: - destructive (Bombs, missiles, mines, artillery) and - disruptive (electromagnetic jamming and electromagnetic deception, expendables i.e., chaff, flares, and decoys). Functional characteristics of sensors: - Intelligence Collection (AN/APG-70, Lantirn, Pave Tack, PDF (ELINT); - Threat detection and identification (AN/APG-70, Lantirn, ESM)	Ground control station: - Observer bay: information collection, analysis, and communication; - Tracking bay: monitoring UAV position; - Remote receiving station: real-time, remote reception & distribution of TV images and intelligence data; - Pilot bay: navigation processes, GPS; flight control; Vehicle characteristics: - Speed, maneuverability, flight profiles, etc. Payload characteristics: high resolution TV & FLIR, radio relays, meteorological sensor, radiac sensor, chemical detection, and COMINT
Inventory Configuration Topography				Map of theater territory with location and: Vehicle types, equipment, and numbers; Weapon types; Sensor types	Map of theater territory with location and type of resources, communication centers, ground stations, portable stations, tracking units, remote receiving stations; Number and configuration of UAVs (Predator, Global Hawk, Darkstar, Pioneer, Hunter, Outrider, Gnat 750, Tiltrotor UAVs), their equipment, weapons, sensors

FIGURE 4.10 (*Continued*)

Abstraction–decomposition space of a military command-and-control system for SEAD missions. (Adapted from Rasmussen, J., 1998, *Ecological interface design for complex systems: An example: SEAD–UAV systems* (AFRL-HE-WP-TR-1999-0011), Air Force Research Laboratory, Human Effectiveness Directorate, Wright-Patterson Air Force Base, OH. With permission.)

Figure 4.11a presents a graphical abstraction hierarchy in a generic form. The nodes within this figure signify constraints. Like all abstraction hierarchies, this one does not indicate the levels of decomposition of the constraints. Figure 4.11b shows one possibility, which is that the constraints fall at the first level of decomposition, but there are many other possibilities. For example, Figure 4.11c shows that the constraints may belong to a combination of levels rather than to the same level. Other possibilities are that the constraints fall at the second level, at the third level, or at a combination of levels different from that displayed in Figure 4.11c.

Whereas Figures 4.11b and 4.11c portray well-formed abstraction hierarchies, Figure 4.11d depicts the kind of abstraction hierarchy that analysts may construct unintentionally if they model a work domain without specifying the decomposition dimension of the representation systematically. In Figure 4.11d, the constraints at each level of abstraction fall at different levels of decomposition rather than within the same level of decomposition. In addition, as highlighted by the nodes drawn with broken lines, some of the constraints that belong in particular cells of the model are missing. In this case, each level of abstraction does not offer a complete representation of a work domain at a particular level of resolution but, instead, an incomplete representation of a work domain at a variety of levels of resolution. Analysts are less likely to produce this kind of model inadvertently if they define the decomposition dimension of the representation methodically—using system decomposition.

System decomposition allows a skeletal abstraction–decomposition space of a system to be developed, so that an overview of the *potential content* of the cells in the model is gained. With this overview, it becomes possible to assess which representations of a work domain are in fact worth creating given the purpose of the analysis. Another advantage of this overview is that it makes it possible to build models with greater comprehensiveness, accuracy, and efficiency.

To construct a skeletal abstraction–decomposition space of a system, it is necessary first to identify the levels of abstraction and decomposition that are appropriate for modeling that system. The abstraction and decomposition dimensions can then be juxtaposed to form a matrix. Consequently, an overview of the potential content of the cells in the matrix, according to their levels of abstraction and decomposition, can be obtained.

Figure 4.12 presents a skeletal abstraction–decomposition space of a home; this is the same home as that shown in Figure 2.7. The vertical and horizontal axes of the matrix portray the abstraction and decomposition dimensions of the model. The italicized text within the cells provides a brief description of their potential content as a function of their levels of abstraction and decomposition in the model.

There are two main ways in which a skeletal abstraction–decomposition space can help analysts to decide which representations of a work domain are the most fruitful to develop. First, because the decomposition hierarchy along the horizontal axis specifies the parts of a system explicitly, it is possible to consider systematically which of those parts are most worth modeling with constraints, given the purpose of the analysis. As discussed previously, the key choices are whether to model all or many parts at each level of decomposition to generate comprehensive work domain representations encompassing several parts of a system (Figure 4.6a) or to model a single part at each level to formulate detailed work domain representations of particular parts of a system (Figure 4.6b). Analysts can also nominate precisely which parts they will model with constraints.

Figure 4.13 utilizes the skeletal abstraction–decomposition space of a home to demonstrate the effects of such decisions on the kinds of representations that are produced. Specifically, this figure shows the parts of the home at each level of decomposition and

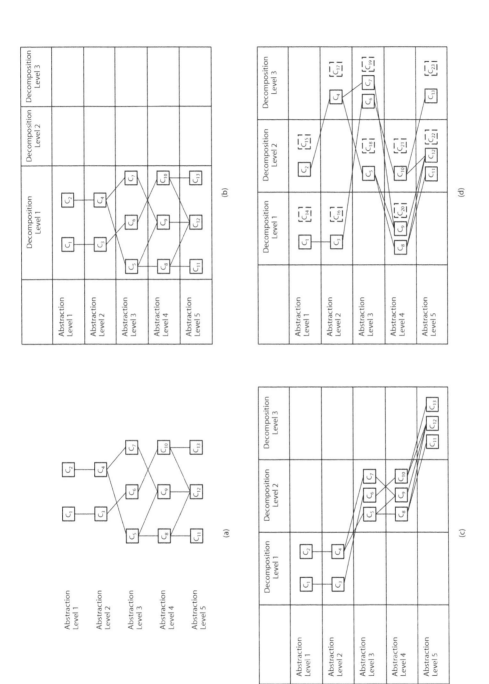

FIGURE 4.11
(a) Graphical abstraction hierarchy; (b) one possibility for the levels of decomposition of the constraints; (c) a second possibility; (d) a poorly formed abstraction hierarchy. C, constraint.

	Whole House	Rooms or Subspaces	Contents
Functional Purposes	*Functional purposes of whole house as a single entity*	*Functional purposes of rooms or subspaces*	*Functional purposes of contents*
Value and Priority Measures	*Value and priority measures of whole house as a single entity*	*Value and priority measures of rooms or subspaces*	*Value and priority measures of contents*
Purpose-related Functions	*Purpose-related functions of whole house as a single entity*	*Purpose-related functions of rooms or subspaces*	*Purpose-related functions of contents*
Object-related Processes	*Functional capabilities or limitations of whole house as a single entity*	*Functional capabilities or limitations of rooms or subspaces*	*Functional capabilities or limitations of contents*
Physical Objects	*Physical form of whole house as a single entity*	*Physical form of rooms or subspaces*	*Physical form of contents*

FIGURE 4.12
Skeletal abstraction–decomposition space of a home.

	Whole House	Rooms or Subspaces	Contents
	Whole house	Kitchen, Dining room, Lounge room, Main bedroom, Guest bedroom, Bathroom, Study, Workshop	Television, Dishwasher, Refrigerator, Stove, Utensils, Compact Disc Player, Armchair, Coffee Table, Shower, Bath, Toothbrush, Towel, Telephone, Computer
Functional Purposes	Functional purposes of _whole house_ as a single entity	Functional purposes of _nominated rooms or subspaces_	Functional purposes of _nominated contents_
Value and Priority Measures	Value and priority measures of _whole house_ as a single entity	Value and priority measures of _nominated rooms or subspaces_	Value and priority measures of _nominated contents_
Purpose-related Functions	Purpose-related functions of _whole house_ as a single entity	Purpose-related functions of _nominated rooms or subspaces_	Purpose-related functions of _nominated contents_
Object-related Processes	Functional capabilities or limitations of whole house as a single entity	Functional capabilities or limitations of _nominated rooms or subspaces_	Functional capabilities or limitations of _nominated contents_
Physical Objects	Physical form of _whole house_ as a single entity	Physical form of _nominated rooms or subspaces_	Physical form of _nominated contents_

FIGURE 4.13

Skeletal abstraction–decomposition space of a home illustrating the effects of decisions about which parts of a system to model with constraints.

indicates what the content of the cells should be in view of the decision to model all or many of the parts at each level. This decision was taken so that the model of a home would provide a variety of examples from different parts of this system, thereby allowing an assortment of material on work domain analysis to be illustrated, rather than detailed examples that are restricted to just a few parts.

Second, a skeletal abstraction–decomposition space allows analysts to assess which cells of the model should be populated with constraints or, in effect, whether full or partial abstraction hierarchies should be composed for the nominated parts at each level of decomposition (Figures 4.6a and 4.6b). In Chapter 2, it was pointed out that, although it is possible to formulate representations for every cell in a model, this is seldom a productive or efficient exercise (Miller and Vicente, 1998). Instead, it is advisable that analysts determine which cells to target by examining the value of the information that each cell offers for the purpose of an analysis. By revealing the potential content of the cells in a model, a skeletal abstraction–decomposition space provides a basis for making informed judgments on this subject.

Figure 4.14 shows the consequences of such decisions on the nature of the model that is created using the skeletal abstraction–decomposition space of a home. The shading in this figure signifies the decision to populate five cells around the diagonal of the model. These cells were chosen because they offered information most relevant to the ideas in this book.

By revealing the potential content of the cells in this way, a skeletal abstraction–decomposition space is also helpful for producing models with greater comprehensiveness, accuracy, and efficiency. Analysts are less likely to miss constraints that belong in the selected cells of a model. For example, having identified the potential content of the cell at the object-related processes level of abstraction and the contents level of decomposition in the skeletal abstraction–decomposition space of a home (Figure 4.14), constraints relevant to the parts of the home associated with that cell are less likely to be omitted inadvertently. Analysts are also less likely to confuse constraints that belong in one cell of a model as belonging to another cell of the model. As a case in point, the probability of mistaking a home's functional purposes at the rooms or subspaces level of decomposition (e.g., provision of meals and beverages) to be the functional purposes of the entire home, so that this constraint is represented in the cell at the coarser level of decomposition, is reduced. In addition, analysts are less likely to mix constraints that belong in different cells of a model within a single cell of the model. For instance, there is less chance of muddling a home's purpose-related functions at the rooms or subspaces level of decomposition (e.g., *rest, personal care and grooming*) with those at the contents level (e.g., washing) and hence representing both types of constraints within a single cell of the model. Analysts are also likely to be quicker, or more efficient, at judging whether a constraint (e.g., provision of meals and beverages, *rest, personal care and grooming*, washing) is relevant to the selected cells of a model.

Two other points are worth noting. The first is that the development of a skeletal abstraction–decomposition space of a system does not preclude the final model from being presented in the form of a graphical or tabular abstraction hierarchy. The second is that, if necessary, it is possible to construct a skeletal abstraction–decomposition space of a system by studying just a few of its parts, as opposed to analyzing all or many of its parts. For example, by observing a small number of the parts of a home, it is possible to establish that a decomposition hierarchy for this system may comprise three levels, specifically whole house, rooms or subspaces, and contents. Despite such a limited method for obtaining an overview of the potential content of the cells in a model, analysts are more likely to develop *efficiently* a *well-formed* model that comprises the most *meaningful* or *useful* representations of a system.

	Whole House	**Rooms or Subspaces**	**Contents**
	Whole house	Kitchen, Dining room, Lounge room, Main bedroom, Guest bedroom, Bathroom, Study, Workshop	Dishwasher, Refrigerator, Stove, Utensils, Television, Compact Disc Player, Armchair, Coffee Table, Shower, Bath, Toothbrush, Towel, Telephone, Computer
Functional Purposes	*Functional purposes of whole house as a single entity*	*Functional purposes of nominated rooms or subspaces*	*Functional purposes of nominated contents*
Value and Priority Measures	*Value and priority measures of whole house as a single entity*	*Value and priority measures of nominated rooms or subspaces*	*Value and priority measures of nominated contents*
Purpose-related Functions	*Purpose-related functions of whole house as a single entity*	*Purpose-related functions of nominated rooms or subspaces*	*Purpose-related functions of nominated contents*
Object-related Processes	*Functional capabilities or limitations of whole house as a single entity*	*Functional capabilities or limitations of nominated rooms or subspaces*	*Functional capabilities or limitations of nominated contents*
Physical Objects	*Physical form of whole house as a single entity*	*Physical form of nominated rooms or subspaces*	*Physical form of nominated contents*

FIGURE 4.14

Skeletal abstraction–decomposition space of a home highlighting the consequences of decisions about which cells of a model to populate with constraints.

Summary

In the past, the decomposition dimension of a work domain model has received much less attention than the abstraction dimension. In this chapter, existing models of assorted systems were reviewed to determine whether there are patterns in the number, types, and labels of the levels of decomposition that analysts can take advantage of in creating new models. It was found that, although many models have a decomposition dimension with either a physical or a conceptual basis, there is no common set of levels of decomposition for characterizing a range of systems. For this reason, the guidelines for work domain analysis in Chapter 8 present a strategy for defining the decomposition dimension of a model without making any assumptions about its characteristics.

Regardless of the number, types, and labels of the levels of decomposition in a work domain model, analysts have adopted two approaches for formulating representations of a system at different levels of detail. Constraint decomposition involves decomposing or aggregating the constraints in a model into more or less detailed representations of a work domain. The standard approach, system decomposition, involves decomposing a system into parts, or aggregating the parts of a system into wholes, to produce a decomposition hierarchy. After that, work domain representations are constructed for the parts at each level using full or partial abstraction hierarchies. Analysts can model either all or many parts at each level to generate comprehensive representations covering several parts of a system (first variation) or a single part at each level to formulate detailed representations of particular parts of a system (second variation).

System decomposition has a significant advantage over constraint decomposition because it allows analysts to construct a skeletal abstraction–decomposition space of a system and thus gain an overview of the potential content of the cells in the model. With this overview, analysts can assess which representations of a work domain are the most meaningful or useful to create given the purpose of the analysis. They can also craft models with greater comprehensiveness, accuracy, and efficiency. For these reasons, it is recommended that analysts do not overlook a methodical analysis of the decomposition dimension of a work domain model—using system decomposition—even when an abstraction hierarchy seems sufficient for an application.

5

Structural Means–Ends, Part–Whole, and Topological Relations

OVERVIEW Structural means–ends and part–whole relations are the principal hierarchical links underlying the abstraction–decomposition space. Sometimes, topological relations, which are introduced in this chapter, are also portrayed in a work domain model. Understanding these relationships sufficiently well to differentiate them from other types of links that should be excluded from this kind of representation can be challenging. Yet, this is essential for composing a sound model. After explaining the distinctions between different types of associations, including those that do and do not belong in a work domain model, this chapter describes structural means–ends, part–whole, and topological relations in more depth.

Types of Relations

This section differentiates structural means–ends, part–whole, and topological relations, which do belong in a work domain model, from action means–ends and is–a relations, which should be excluded from this kind of representation. The distinctions between these types of links are demonstrated with examples from the model of a home (Figure 2.7). These examples are sometimes simplified so that the different associations can be compared more easily.

Structural means–ends relations define the abstraction dimension of a work domain model. As pointed out in Chapter 2, nodes at lower levels of abstraction identify the structural means for achieving the ends at higher levels, whereas nodes at higher levels specify the ends that can be attained by the structural means at lower levels. Figure 5.1a provides two examples of structural means–ends relations in the context of a home. The links in this figure signify such relationships because both a *bed* and an *armchair* are structural means for achieving the end of *rest*.

Structural means–ends relations should be distinguished from action means–ends relations, which have no place in a work domain model (Vicente, 1999). Action means–ends relations describe the tasks or activities for attaining an end. Hence, in an action means–ends hierarchy, nodes at lower levels specify the action means for achieving the ends at higher levels. Conversely, nodes at higher levels identify the ends that can be obtained by the action means at lower levels. The connections between the nodes in Figure 5.1b denote this type of relationship because both *lie on bed* and *sit in armchair* are action means for attaining the end of *rest*.

Action means–ends relations should not be included in a work domain model because this kind of model represents the functional structure of actors' *environments* rather than

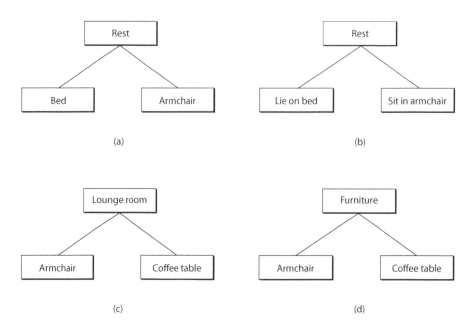

FIGURE 5.1
Types of hierarchical relationships: (a) structural means–ends relations; (b) action means–ends relations; (c) part–whole relations; (d) is–a relations.

their *behavior.* In the context of a home, a *bed* and an *armchair* are structural properties of the environment that actors may utilize to accomplish *rest* (Figure 5.1a). In contrast, *lie on bed* and *sit in armchair* are behaviors that actors may perform to achieve *rest* (Figure 5.1b). The structural properties of the environment shape actors' behavior. Therefore, depending on whether a *bed* or an *armchair* is used for *rest*, the behavior of actors may be different.

Part–whole relations characterize the decomposition dimension of a work domain model. In Chapter 2, it was explained that nodes at lower levels of decomposition are the functional parts of entities at higher levels. Conversely, nodes at higher levels are the functional wholes of entities at lower levels. Two examples of part–whole relations are shown in Figure 5.1c. The links in this figure are of this type because both an *armchair* and a *coffee table* are functional parts of a *lounge room*, while a *lounge room* is the functional whole of the entities at the level below.

Is–a relations are another type of association that should be excluded from a work domain model (Vicente, 1999). These relationships define a classification hierarchy. Nodes at lower levels are exemplars of the nodes at higher levels. Alternatively, nodes at higher levels are the superordinate categories of those at lower levels. The links in Figure 5.1d denote such connections because both an *armchair* and a *coffee table* are exemplars of *furniture*, whereas *furniture* is a superordinate category to which an *armchair* and a *coffee table* belong.

Finally, as well as structural means–ends and part–whole relations, topological relations may be represented in a work domain model. Unlike the preceding types of links, topological relations are not hierarchical. That is, they do not connect nodes that fall at the different levels of a hierarchy. Instead, they connect nodes *within* the cells of the abstraction–decomposition space or *within* the levels of the abstraction hierarchy. Topological relations, therefore, may be contrasted with structural means–ends and part–whole relations, which link nodes across the different levels of the abstraction or decomposition hierarchy, respectively (Figure 5.2a).

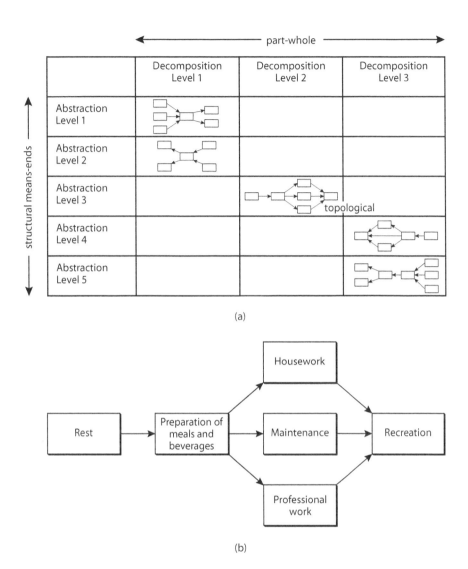

(a)

(b)

FIGURE 5.2
(a) Unlike structural means–ends and part–whole relations, topological relations connect nodes that fall *within* the cells of the abstraction–decomposition space; (b) example of topological relations at the purpose-related functions level of abstraction and the rooms or subspaces level of decomposition in the model of a home.

The specific nature of the topological relations in a work domain model not only depends on the kind of system but also varies as a function of the levels of abstraction and decomposition in the model. Figure 5.2b illustrates the concept of topological relations in the context of the home model (Figure 2.7). This figure represents one possibility for the flow of people between the nodes at the purpose-related functions level of abstraction and the rooms or subspaces level of decomposition. The links in this figure are topological relations because they relate nodes that belong within a cell of the abstraction–decomposition space. In this case, the associations between the nodes are based on the flow of people through the system, but, depending on the kind of system and the levels of abstraction and decomposition in the model, the nature of the topological relations may be different.

Structural Means–Ends Relations

Structural means–ends relations are one of the two hierarchical relationships that are fundamental to a work domain model. This section provides a few more examples of such associations from models of two systems other than a home. Following that, some finer points about what they reveal in a work domain model are addressed.

Figure 5.3 presents a model of a microworld simulation of a pasteurization plant (Reising and Sanderson, 2002b). Some of the nodes and links in this figure have been accentuated for the purposes of this discussion. Moving from the lowest to the highest level of abstraction, the highlighted areas show that *milk circulation* is a means by which *feedstock flow* and *heat exchange* can be achieved. *Heat exchange* provides a means by which *regulatory compliance* and *energy balance* can be realized. In the same way, *energy balance* is a means by which *temperature demand* can be met.

Figure 5.4 depicts a portion of an abstraction hierarchy of the human body in a resting state, when it is under anesthetic (Watson and Sanderson, 2007). As before, the highlighted nodes and links show that the *heart* provides a means for *pumping blood to the lungs* and *pumping blood to the body*. *Pumping blood to the body* is a means by which *circulate oxygenated blood to the body* can be attained. *Circulate oxygenated blood to the body*, in turn, provides a means by which *maintain intracellular function* and *maintain cellular function* can be realized. Last, both *maintain intracellular function* and *maintain cellular function* are means for achieving *homeostasis*.

Alternate Combinations of Means–Ends Relations

A work domain model represents the alternative means, the combinations of means, or the "alternate combinations" of means to achieve an end (Rasmussen et al., 1994, p. 40). For example, in the model of a home (Figure 2.7), *rest* can be achieved with the *sleeping capacity* of a *bed* or the *seating capacity* of an *armchair* (Figure 5.5a). This represents the alternative means that can be used to attain an end, or an OR relationship. In contrast, both the *seating capacity* of an *armchair* and the *reception and audiovisual capacity* of a *television* may be required for *recreation* (Figure 5.5b). This represents the combination of means that can be used to obtain an end, or an AND relationship. Similarly, both the *seating capacity* of an *armchair* and the *audio playback capacity* of a *compact disc player* may be required for *recreation* (Figure 5.5c). This, too, represents an AND relationship. However, either combination—the *seating capacity* of an *armchair* AND the *reception and audiovisual capacity* of a *television* OR the *seating capacity* of an *armchair* AND the *audio playback capacity* of a *compact disc player*—may be used for *recreation* (Figure 5.5d). This represents the alternate combinations of means that can be used to realize an end. By representing the alternative means, the combinations of means, and the alternate combinations of means to achieve an end, a work domain model captures all of the possibilities for action that are available to actors.

Which of these relationships is represented by a link in a model is usually not indicated. There are several reasons why this may be the situation. One is that it may be unnecessary for an application. Another is that the resulting representation would be too complex. In this case, if it is useful to do so, the specific nature of the links may be recorded in another form, for instance, in a supplementary document. A third reason is that it is generally not possible to anticipate, and thus depict, all of the means–ends relations in a model, as the following section explains.

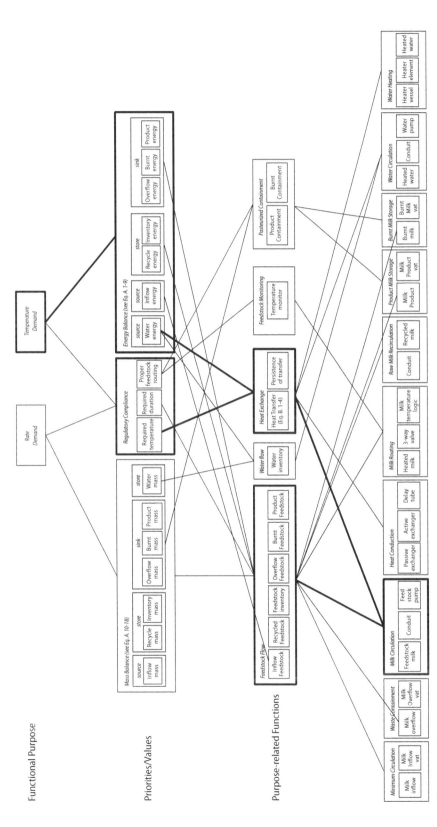

FIGURE 5.3

Model of a microworld simulation of a pasteurization plant. The shaded nodes are at a higher level of decomposition relative to the clear ones they enclose. The physical objects level of abstraction is not relevant because the model is of a computer simulation. (Adapted from Reising, D. V. C., and Sanderson, P. M., 2002, Work domain analysis and sensors II: Pasteurizer II case study, *International Journal of Human-Computer Studies*, 56(6), 597–637. Copyright © 2002. With permission from Elsevier.)

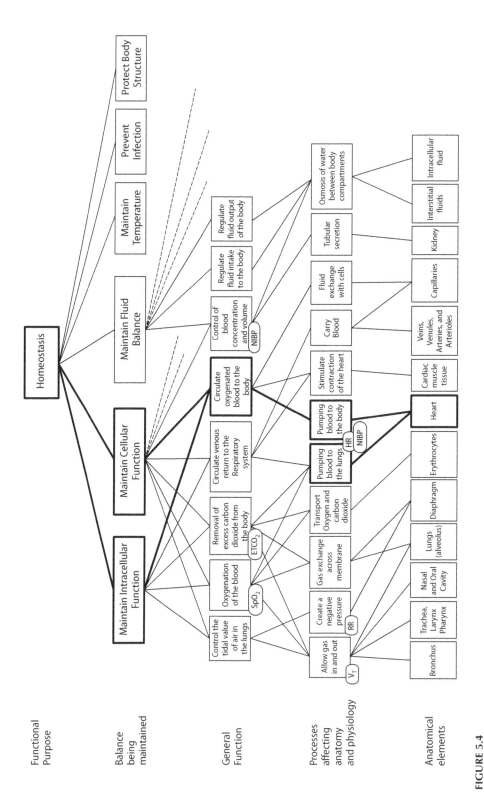

FIGURE 5.4
Portion of an abstraction hierarchy of the human body in a resting state. (Adapted from Watson, M. O., and Sanderson, P. M., 2007, Designing for attention with sound: Challenges and extensions to ecological interface design, *Human Factors, 49*(2), 331–346. Copyright © 2007 by the Human Factors and Ergonomics Society. All rights reserved.)

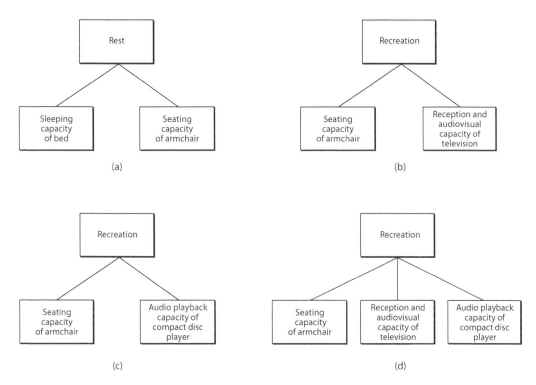

FIGURE 5.5
(a) Alternative means to achieve an end; (b) one combination of means to achieve an end; (c) a second combination of means to achieve an end; (d) alternate combinations of means to achieve an end.

Instantiations of Means–Ends Relations

A graphical abstraction–decomposition space or graphical abstraction hierarchy portrays the means–ends relations in a model explicitly by drawing links between the constraints.* Typically, these links do not represent all of the means–ends relations that are possible but particular instantiations of such relationships. One instantiation that might be shown is those means–ends relations that are deliberately designed or engineered into a system. A model of a microworld simulation of a pasteurization plant in Figure 5.3 is an example of this case (D. V. C. Reising, personal communication, March 26, 2008). Another potential instantiation is all of the means–ends relations that are known or can be anticipated by subject matter experts. An abstraction hierarchy of the human body at rest, which was also discussed previously (Figure 5.4), is an example of this type of model (M. O. Watson, personal communication, March 27, 2008). A third kind of instantiation is those means–ends relations that are evident in a particular situation or set of situations. This instantiation is illustrated by a representation of an emergency ambulance dispatch management system (Figure 5.6), which appeared in Hajdukiewicz et al. (1999). Specifically, Figure 5.6a describes the constraints in this model, and Figure 5.6b indicates the means–ends relations that are apparent in a certain situation. Any combination of these three instantiations of means–ends relations might also be depicted in a model.

---

* Chapter 2 discusses formats for presenting work domain models.

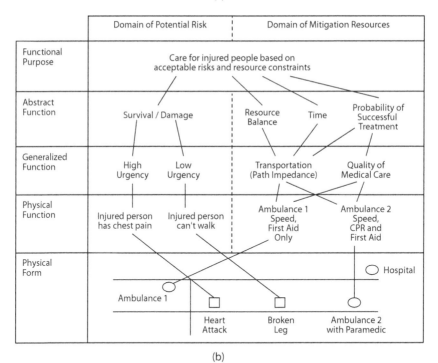

	Domain of Potential Risk	Domain of Mitigation Resources
Functional Purpose	Goal: Maximize emergency health care response given the level of risk that is acceptable; Constraints: The resources and abilities associated with the emergency response system	
Abstract Function	Priority criteria and given level of acceptable risk (e.g., survival, damages, public opinion, costs, rules and regulations)	Time, resource balance within region, probability of successful treatment
Generalized Function	Urgency of injury	Transportation, quality of care
Physical Function	Functional impairment, consequences of injury	Functional capability of resources (e.g., qualifications of medical personnel, speed of ambulance)
Physical Form	Location, number of people injured, type of injury	Locations, inventories, availability of various response resources (e.g., helicopters, ambulances, stations, hospitals, command posts, paramedic officers), roads, weather, terrain, traffic

(a)

(b)

FIGURE 5.6

Model of an emergency ambulance dispatch management system showing (a) constraints (From Rasmussen, J., Pedersen, O. M., and Grønberg, C. D., 1987, *Evaluation of the use of advanced information technology (expert systems) for data base system development and emergency management in non-nuclear industries* (Risø-M-2639), Risø National Laboratory, Roskilde, Denmark. With permission; as adapted by Hajdukiewicz et al., 1999) and (b) means–ends relations that are evident in a particular situation. (From Hajdukiewicz, J. R., Burns, C. M., Vicente, K. J., and Eggleston, R. G., 1999, Work domain analysis for intentional systems, *Proceedings of the Human Factors and Ergonomics Society 43rd Annual Meeting*, 333–337. Copyright © 1999 by the Human Factors and Ergonomics Society. All rights reserved.)

Generally, none of these instantiations or combinations of instantiations will represent all of the means–ends relations that are possible. Actors may adapt their behavior to suit variations in circumstances or their individual preferences in ways that are not deliberately designed or engineered into a system, are not known or anticipated by subject matter experts, or are not evident in particular situations. Actors may also formulate innovative behaviors to deal with novel or unanticipated events; by definition, these behaviors cannot be specified or observed prior to their occurrence. As a result, actors may 'discover' or 'reveal' means–ends relations that are not captured by any instantiation or combination of instantiations of means–ends relations.

A model, then, is not necessarily incomplete if a new means–ends link could be drawn between existing constraints. Instead, an incomplete model is one to which new constraints must be added. As highlighted by Jamieson (personal communication, November 23, 2010), though, the process of identifying as many means–ends relations in a model as possible is helpful for checking the comprehensiveness and accuracy of the constraints.

Weighting of Means–Ends Relations

The means–ends relations in a work domain model are not weighted in terms of importance. This is because their priority is dependent on the situation. In the case of the model of a home (Figure 2.7), a *shower* and its *showering capacity* may have greater priority on weekday mornings when inhabitants have to go to work, whereas a *bath* and its *bathing capacity* may be of greater importance on weekends when time is not such a pressing concern. Moreover, it is difficult to predict the importance of links in unanticipated situations. A link that is insignificant in most situations may in fact have the highest priority during an unforeseen event.

Part–Whole Relations

As well as structural means–ends relations, part–whole relations are fundamental to a work domain model. In this section, models of two systems other than a home are used to provide some more examples of this type of association. An apparent duplication that is sometimes evident between the abstraction and decomposition hierarchies of a model is also discussed.

Figure 5.7 presents a decomposition hierarchy of an acetylene hydrogenation reactor in a petrochemical plant (Miller and Vicente, 1998). The first level of decomposition, system, comprises the acetylene hydrogenation reactor. The next level is defined by subsystems such as feed heating unit and cooling unit, which are functional parts of the acetylene hydrogenation reactor. The last level consists of components such as reactor feed preheater and reactor heat cross exchanger. These components, which are depicted diagrammatically, are the functional parts of the subsystems at the level above.

Figure 5.8 presents a decomposition hierarchy of the cardiovascular system of the human body[*] (Hajdukiewicz, 1998). The first level of decomposition comprises the whole body, and the second level is defined by the cardiovascular system, which is a functional

---

[*] See Figures 3.8a and 3.8b, respectively, for the corresponding abstraction–decomposition spaces of the human body and its cardiovascular system.

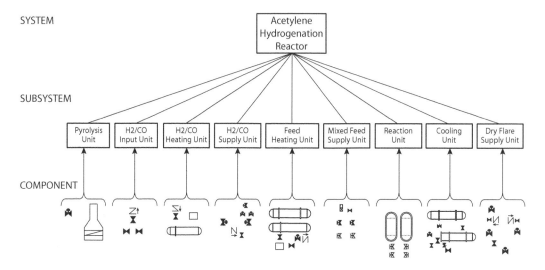

FIGURE 5.7
Decomposition hierarchy of an acetylene hydrogenation reactor in a petrochemical plant. (Adapted from Miller, C. A., and Vicente, K. J., 1998, *Abstraction decomposition space analysis for NOVA's E1 acetylene hydrogenation reactor* (CEL 98-09), Cognitive Engineering Laboratory, Toronto, Ontario, Canada. With permission.)

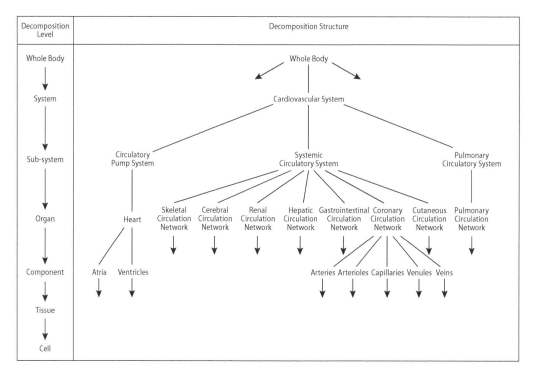

FIGURE 5.8
Decomposition hierarchy of the cardiovascular system of the human body. (Adapted from Hajdukiewicz, J. R., 1998, *Development of a structured approach for patient monitoring in the operating room,* unpublished master's thesis, University of Toronto, Ontario, Canada. With permission of the author.)

part of the human body. The third level includes subsystems such as the circulatory pump system and the systemic circulatory system, which are functional parts of the cardiovascular system. The fourth level comprises organs, including the heart and the skeletal circulation network, which are functional parts of the subsystems. The fifth level consists of components such as the atria and the ventricles, which are functional parts of the organs. According to Hajdukiewicz, the content of the sixth (tissue) and seventh (cell) levels was not represented in this figure because of its complexity.

Sometimes, an apparent duplication may be evident between the abstraction and decomposition hierarchies of a model. This may occur whenever the decomposition dimension of a model has a physical basis* and the names of physical objects are listed at one of the levels of abstraction in the representation. For example, the model of a home (Figure 2.7) is characterized by physical decomposition and the names of physical objects are noted at the lowest level of abstraction. Consequently, as indicated by the shading in Figure 5.9, the representations at the contents level of the decomposition hierarchy and the physical objects level of the abstraction hierarchy are the same. The names of physical objects at the lowest level of abstraction in this model, though, are placeholders for more detailed descriptions of their physical forms. As noted in Chapter 3, the names of physical objects are often listed in a model in place of descriptions of their physical forms or their functional processes, unless such descriptions are necessary for an application. This leads to an *apparent*, rather than actual, duplication between the abstraction and decomposition hierarchies of a model. The duplication is only apparent because the names of physical objects in the decomposition hierarchy stand for the physical objects themselves, whereas at the lowest two levels of the abstraction hierarchy, the names of physical objects are substitutes for their physical forms or functional processes. Such an apparent duplication may also be evident in the case of conceptual decomposition, specifically if one or more levels of decomposition are characterized by physical objects.

Topological Relations

In addition to structural means–ends and part–whole relations, topological relations are sometimes represented in a work domain model. As highlighted previously, topological relations refer to connections between nodes located within the cells of the abstraction–decomposition space or within the levels of the abstraction hierarchy. This section illustrates that the nature of these connections may vary as a function of the type of system as well as the levels of abstraction and decomposition in the model. Differences in the nature of the topological relations in models of causal versus intentional systems are emphasized.

Prior to that, it is worth noting that topological relations have generally been represented in models of causal systems that were produced to support the design of interfaces. In this context, such associations, like structural means–ends and part–whole relations, reveal how information might be organized or integrated on a display. Burns and Hajdukiewicz (2004) provide a number of examples of how the topological relations in a model have shaped the design of interfaces for various systems. (They refer to models portraying such relationships as causal models.)

---

* Both physical and conceptual decomposition are explained in Chapter 4.

	Whole House	Rooms or Subspaces	Contents
	Whole house	Dining room, Kitchen, Lounge room, Main bedroom, Bathroom, Guest bedroom, Study, Workshop	Dishwasher, Refrigerator, Stove, Utensils; Television, Compact Disc Player, Armchair, Coffee Table; Shower, Bath, Toothbrush, Towel; Telephone, Computer
Functional Purposes	Well-being, Environmental protection, Residential laws and regulations etc.		
Value and Priority Measures	Total income – total expenses = savings, Time, Health, Hygiene, Pleasure, Conservation of natural resources etc.		
Purpose-related Functions		Preparation of meals and beverages, Dining, Recreation, Rest, Exercise, Personal care and grooming, Housework, Maintenance, Professional work etc.	
Object-related Processes			Washing capacity, Cooling capacity, Heating capacity, Food and beverage handling capacity, Serving and seating capacity, Reception and audiovisual capacity, Audio playback capacity, Seating capacity, Serving and exhibiting capacity, Sleeping capacity, Showering capacity, Bathing capacity, Brushing capacity, Drying capacity, Transmission and audiovisual capacity, Data handling capacity etc.
Physical Objects			Dishwasher, Refrigerator, Stove, Utensils, Table, Television, Compact disc player, Armchair, Coffee table, Bed, Shower, Bath, Toothbrush, Towel, Telephone, Computer etc.

FIGURE 5.9

Apparent duplication between the abstraction and decomposition hierarchies in the model of a home.

Figure 5.10 shows the topological relations in an abstraction–decomposition space of DURESS II[*] (Bisantz and Vicente, 1994), a thermalhydraulic microworld simulation that may be classified as a causal system. In the cell at the abstract function and subsystem levels, the topological relations represent causal flows of *mass* and *energy* through the subsystems of DURESS II. In contrast, in the cell at the generalized function and subsystem levels, the links denote causal flows of *water* and *heat* through the same subsystems. In the cell at the generalized function and component levels, the links also represent flows of *water* and *heat*, but through the components of DURESS II. Similarly, in the next two cells, the topological relations signify connections between the components. Specifically, at the physical function and component levels, the links indicate *physical connections*, whereas at the physical form and component levels, the topological relations are *spatial relationships*, although these are not depicted in Figure 5.10.

Stated differently, at the abstract function level, the topological relations represent causal flows of mass or energy, whereas at the generalized function level, they indicate causal flows of water or heat. At the physical function and physical form levels, the topological relations signify physical connections and spatial relationships, respectively. Depending on the levels of decomposition at which these associations are modeled, the causal flows, physical connections, or spatial relationships are portrayed between the subsystems or components of DURESS II.

Some other work domain models of process control systems similar to DURESS II also depict topological relations. These include models of a feedwater subsystem of a nuclear power plant (Dinadis and Vicente, 1996), a simulation of a coal-fired power plant (Burns, 1998), and a simulation of a fluid catalytic cracking unit in a petrochemical refinery[†] (Jamieson, 1998; Jamieson and Vicente, 2001). The nature of the topological relations at the abstract function, generalized function, and physical function levels of abstraction in these models is similar to that in the model of DURESS II. One difference is that, in the model of the fluid catalytic cracking unit, the links at the generalized function level[‡] represent flows of water and commodities (e.g., catalyst, feed products) rather than water and heat. As for DURESS II, the levels of decomposition in these models determine the levels of detail at which the topological relations are represented. The physical form level was not populated in these models because the models are of computer simulations.

An example of topological relations in a causal system other than a process control system is provided by Hajdukiewicz's (1998) representation of the cardiovascular system of the human body. This model has five levels of abstraction (purposes, balances, processes, physiology, and anatomy) and four levels of decomposition (system, subsystem, organ, and component).[§] Figure 5.11 portrays the topological relations at the balances and processes levels of abstraction and the system, subsystem, and organ levels of decomposition. At the balances level, the links denote flows of *mass*, whereas at the processes level, the links signify flows of *fluids*. Depending on the levels of decomposition at which the topological relations are modeled, the flows of mass or fluids are represented at the level of the entire cardiovascular system, its subsystems, or its organs. In the sense that the topological relations at the second and third levels of abstraction in this model represent causal flows,

---

[*] See Figure 2.15 for the means–ends relations in the model of DURESS II.

[†] See Figures 4.1a and 4.1b for an overview of the abstraction–decomposition space and the decomposition hierarchy of a simulation of a fluid catalytic cracking unit in a petrochemical refinery.

[‡] The topological relations at the generalized function level in the model of the fluid catalytic cracking unit are shown in Figures 4.8 and 4.9.

[§] Figure 3.8b presents the abstraction–decomposition space of the cardiovascular system of the human body, and Figure 5.8 displays the decomposition hierarchy of this system.

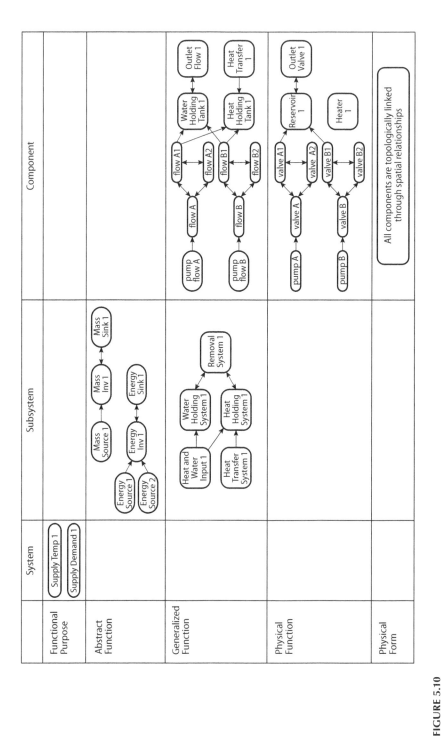

FIGURE 5.10
Topological relations in an abstraction–decomposition space of DURESS II, a thermalhydraulic microworld simulation. (From Bisantz, A. M., and Vicente, K. J., 1994, Making the abstraction hierarchy concrete, *International Journal of Human-Computer Studies, 40,* 83–117. Copyright 1994. With permission from Elsevier; as adapted by Vicente, 1999.)

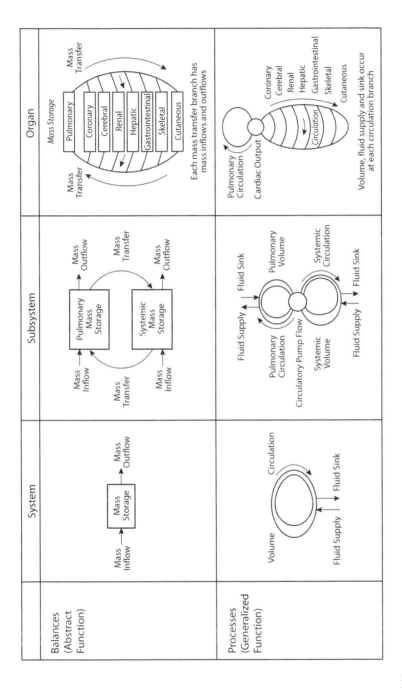

FIGURE 5.11
Topological relations in an abstraction–decomposition space of the cardiovascular system of the human body. (Adapted from Hajdukiewicz, J. R., 1998, *Development of a structured approach for patient monitoring in the operating room*, unpublished master's thesis, University of Toronto, Ontario, Canada. With permission of the author.)

they are similar to those at the second and third levels of abstraction in the models of the process control systems discussed previously. Topological relations were not portrayed at the other levels of abstraction in this model.

Few examples are available of topological relations in models of intentional systems. Kinsley et al. (1994) provide a reasonably detailed discussion of such associations in advanced manufacturing systems, which have a greater degree of intentional constraints compared with the systems referred to earlier in this section. Although Kinsley et al. do not present a graphical or tabular representation of an advanced manufacturing system, they do describe the five levels of abstraction in their model (i.e., system purpose, abstract function, generalized function, physical function, and physical form) and the nature of the topological relations at the last four levels.

At the abstract function level, they identify four kinds of topological flows: the mass topology, which represents the flow of material (pallets) through the system; the delay topology, which signifies the time delays associated with each pallet as it flows along its course; the value topology, which indicates the value added to each pallet as it flows through the system; and the priority topology, which represents the priority associated with each pallet as it flows along its route. At the generalized function level, the topological relations describe the flow of pallets according to the system's general functions (e.g., loading, unloading, setup, machining, assembly, inspection, material handling, scheduling), and at the physical function level, they describe the flow of pallets in terms of the system's processes (e.g., milling, cutting, grinding). At the physical form level, the topological relations represent the physical connections between the system's components, highlighting all of the possible physical paths that a pallet can take through the system.

Kinsley et al. (1994) observe some key differences between the nature of topological relations in highly causal systems, such as process control systems, and systems with more intentionality, such as advanced manufacturing systems. In highly causal systems, topological relations are determined by physical or natural laws and are hardwired into the system. Workers have little discretion in determining the flows in these systems. In contrast, in systems with greater intentionality, there are multiple possibilities for topological relations, and workers have significant flexibility in choosing the flows on any occasion. For instance, in an advanced manufacturing system, workers can select the order in which a machine should process jobs or decide which of several routes a part to be manufactured should take through the system.

The home in Figure 2.7 is characterized by a much greater degree of intentionality than an advanced manufacturing system. In the home, there are multiple, if not infinite, possibilities for topological relations. Figure 5.2b shows one possibility for the flow of people at the purpose-related functions level of abstraction and rooms or subspaces level of decomposition, but there are many other possibilities. Actors also have the flexibility to determine the topological flows in any situation. The same observations apply to the other cells of this model. Hence, this example highlights that highly intentional systems are likely to be characterized by multiple, if not infinite, possibilities for topological relations.

Summary

Structural means–ends and part–whole relations are fundamental to a work domain model, but topological relations may also be depicted in this kind of representation. Unlike structural means–ends and part–whole relations, which are hierarchical as they connect nodes between the different levels of the abstraction or decomposition hierarchy, respectively, topological relations connect nodes within the cells of the abstraction–decomposition space or levels of the abstraction hierarchy. Distinguishing these types of links from other kinds of associations that do not belong in a work domain model, such as action means–ends and is–a relations, is not always easy, but its importance for constructing a sound model cannot be overstated.

As well as illustrating the different types of links with examples from various systems, this chapter made a number of important observations about structural means–ends, part–whole, and topological relations. Specifically, these observations are as follows:

1. A work domain model represents the alternative means, the combinations of means, and the alternate combinations of means to achieve an end.

2. The means–ends relations portrayed graphically in a model are instantiations of means–ends relations and not all of the means–ends relations that are possible.

3. The means–ends relations in a work domain model are not weighted in terms of importance because their priorities are dependent on the situation.

4. An apparent duplication between the abstraction and decomposition hierarchies of a model may be evident when the decomposition dimension has a physical basis, or a conceptual basis with one or more levels characterized by physical objects, and one of the levels of abstraction lists the names of physical objects (perhaps as substitutes for their physical forms or functional processes).

5. The nature of the topological relations in a model may vary as a function of a system's characteristics, including whether it is causal or intentional, as well as the levels of abstraction and decomposition in the model.

The last three chapters dealt with particular features of the abstraction–decomposition space, specifically the abstraction dimension, the decomposition dimension, and the types of relations that constitute this modeling tool. The next two chapters attend to two other topics that are also essential for understanding work domain analysis. Chapter 6 addresses reasons for creating multiple models of the same system, and Chapter 7 examines the relationship between the analysis of activity and work domain analysis as the standard texts can be confusing on this point.

6

Multiple Models

OVERVIEW While in most cases analysts create a single model of a system when performing work domain analysis, there are times when it is useful to construct multiple models. This chapter examines two reasons for developing multiple models of the same system. The first is to depict multiple stakeholders' perspectives of a problem, and the second is to differentiate between multiple facets of a problem. Some brief observations about the relationships between multiple models and decomposition are also made.

Introduction

It is useful to begin with brief examples of the two types of multiple models considered in this chapter before discussing them in detail. The two can be illustrated clearly in relation to the problem of financial management in a home. In the case of the first type, which portrays a problem from the viewpoint of different stakeholders, multiple models may be composed to represent the perspectives of both the inhabitants and their accountants. With respect to the second type, which emphasizes the distinct aspects of a problem, multiple models may be formulated to differentiate the sources of expenses from the sources of income.

I use the term *multiple models* to refer to a series of abstraction–decomposition spaces or abstraction hierarchies of the same system. Multiple abstraction hierarchies are more common, but, as I demonstrate later in this chapter, it is also possible to construct multiple abstraction–decomposition spaces.

Multiple Stakeholders' Perspectives

One reason for assembling multiple models of the same system is to portray different stakeholders' perspectives of a problem. *Stakeholders* may be described as individuals or groups with different views of a particular problem. Stakeholders generally have something to gain or lose with respect to the problem under consideration, which means that they usually have to interact or coordinate their activities to achieve their respective objectives. The views or perspectives of stakeholders may be described as their "object worlds" (Bucciarelli, 1988, p. 162; Rasmussen, 1990, p. 13).

The term *object world* was used by Bucciarelli (1988) to describe his observations during a study of engineering design. Bucciarelli found that the participants in a design project had different perspectives of the object or artifact under design. The participants' object worlds depended mainly on their technical specializations, which focused their attention on particular subsets of attributes of the same artifact. For instance, in a project that involved designing "photovoltaic modules for the direct conversion of solar radiation into electricity," an electrical engineer was concerned with voltage potentials and current flows associated with the design artifact, whereas the project manager was focused on schedules, manpower requirements, and cost trends connected with that artifact (Bucciarelli, 1988, p. 162).

Following Bucciarelli's (1988) study, Rasmussen (1990) developed multiple models of three separate engineering design systems, using a series of abstraction hierarchies for each, to capture the perspectives of different stakeholders in those systems. As pointed out by Rasmussen et al. (1994), stakeholders' object worlds are usually shared or coupled with each other, to varying degrees, at different levels of abstraction. Consequently, changes or effects relating to one stakeholder can propagate to other stakeholders. It is this coupling between stakeholders that might make it useful to develop multiple models to capture their different viewpoints. A potential benefit is that it becomes possible to introduce designs for one stakeholder without creating unintended side effects on other stakeholders. In addition, designs may be produced that allow stakeholders to coordinate their activities more effectively.

One example of multiple models of stakeholders' perspectives was developed by Burns and Vicente (1995, 2000) for an engineering design system. The stakeholders in this system were all concerned with the design of a control panel for the control room of a nuclear power plant. As indicated in Figure 6.1, the stakeholders included human factors designers, structural design engineers, implementers, customers, and management. Each stakeholder had a different view or perspective of the design of the control panel, as summarized in

	HF Design	Structural Design	Implementers	Customer (Utility)	Upper Management
View	display surface for indicators and controls	physical housing for indicators and controls	something they have to produce	furnishings in their control room	contract completion
Objectives	visibility, operability	strength, stability	feasibility	image, cost	marketshare, on time, within budget
Processes	viewing angles, reach envelopes	seismic testing	⑥ manufacturing processes, shipping, installation, on-site modifications ②	approval process	scheduling, resource allocation
Physical Components	panel dimensions, panel geometry, room configuration ① anthropometric data ⑤	construction materials ③		room dimensions, building dimensions, plant staffing	schedule, personnel, $, resources

FIGURE 6.1
Multiple models of stakeholders' perspectives in an engineering design system; these stakeholders were concerned with the design of a control panel for a nuclear power plant. (From Burns, C. M., and Vicente, K. J., 1995, A framework for describing and understanding interdisciplinary interactions in design, *Proceedings of DIS '95: Symposium on Designing Interactive Systems*, 97–103. Copyright © 1995 ACM, Inc. Adapted by permission; incorporating modifications made by Vicente, 1999.)

the 'View' row of Figure 6.1. Hence, each stakeholder approached this design problem differently. For instance, whereas human factors designers were concerned with creating a control panel that was comfortable and usable, structural design engineers were focused on developing a control panel with a certain strength, rigidity, and lifetime of use. Burns and Vicente (2000) observed that the views of all of the stakeholders were correct, but that each stakeholder only had a partial view of the design problem. In addition, although there were vast differences in the goals of the stakeholders, as reflected at the highest level of abstraction (objectives) in the representation, the stakeholders shared many concerns over the physical details of the design problem, such as the panel dimensions, as denoted by the ellipses at the lowest level of abstraction (physical components).

Using the representation they had constructed, Burns and Vicente (2000) demonstrated that the design activities of one stakeholder tended to force reactionary design activities by other stakeholders. Figure 6.1 depicts one trajectory of events they observed. This trajectory began with the discovery of the size of the hallways in the building where the control panel was to be installed (event 1). When the hallway sizes were compared with the dimensions of the control panel, problems with the shipment and installation of the panel were identified. That is, the panel was too large to be shipped through the hallways of the building (event 2). The panel dimensions were revised (event 3), and the human factors or ergonomics concerns were rechecked (event 4). Then, a decision was made to build the panel in smaller segments instead of altering its overall dimensions (event 5). This decision affected the manufacturing processes for the control panel (event 6). Several other such trajectories of the propagation of the effects of design activities across different stakeholders were observed (Burns and Vicente, 2000; Vicente, 1999).

Benda and Sanderson (1998) also developed multiple models of the perspectives of stakeholders in an engineering design system. Unlike Burns and Vicente (2000), they were concerned with predicting the propagation of the effects of design changes across stakeholders instead of mapping trajectories of events that had already occurred. The stakeholders in this study were concerned with the design of an elevator system. As shown in Figure 6.2, the stakeholders included users (public, maintenance); the elevator firm (operations research, engineering, designers); and decision makers (builder, client/owner, architect). Benda and Sanderson observed that the elevator system was common to all of the stakeholders at the lowest level of abstraction (physical forms and configuration) in their representation.

The trajectories in Figure 6.2 denote a few of Benda and Sanderson's (1998) predictions of how the effects of hypothetical design changes to the elevator system might propagate across stakeholders. These trajectories are initiated with the introduction of buttons for floor selection into the lobbies of a building's ground floor. Members of the public can select their destinations prior to entering the elevator. As a result, waiting times for the public are reduced because the elevator system has earlier information about the number of passengers and their destinations (event 1). However, the floor selection buttons lead to greater design complexity for the elevator system, with more wiring, more buttons, and additional display and space concerns (event 2). The greater design complexity of the elevator system, in turn, can lead to usability problems for the public (event 3).

Another example of multiple models of stakeholders' perspectives is provided by Rasmussen et al. (1990, 1994) in the context of a health care system. Whereas the stakeholders in the engineering design systems were focused on the design of a particular object or artifact, the stakeholders in the health care system were concerned with the treatment of patients in a hospital. Figure 6.3 shows a representation of the health care system (Rasmussen et al., 1994). The main point to note about this representation is that the

	USERS			ELEVATOR FIRM			DECISION MAKERS	
	Public	Maintenance	Operations Research	Engineering	Designers	Builder	Client/Owner	Architect
Goals	Transport	Minimize service times Keep system fully functional	Efficient operation	Build operational system Build innovative system	Create marketable design Create distinguished design	Accommodate lift system Reputation Acquire contracts	Profit Reputation	Appropriate design Reputation
Priorities and Values	Safety Simplicity Privacy No waiting Choice ③	Early warning Quick diagnosis Low error rates Accessibility Maintainability	Reduce wait times Keep queue sizes within capacity limits ... other OR measures ①	Safety Reliability Speed Comfort Novelty	Usability and functionality Aesthetics Maintainability	Cost control Timeliness Reliability Installability	Cost control Customer satisfaction Timeliness Upkeep	Aesthetics Functionality Innovate
Generalized Function	Selection	Predict problems Detect . . . Manage . . . Diagnose . . . Repair problems	Measure demand Identify usage patterns Traffic studies	Mechanical Electronic Electrical Energy-related	Design process Guide users Inform users	Subcontracting Project scheduling and management	Express needs Decide architect Deal with occupants	Identify client needs Design process Supervise installation
Object Related Functions and Processes	Find lifts Call car Enter car Select floor Exit car	Test Solder Screw Lubricate	Sensing Simulating	Lifting Lowering Opening Closing	Provide floor information Provide emergency support information	Install Insert Pound Wire	Clean Speak Provide feedback	Draw Modify
Physical Forms and Configuration	**Elevator System** - Lobby - Doors - Car - Panel ②	**Elevator System** - Lobby - Doors - Car - Panel ② Manuals Diagnostic equipment	**Elevator System** - Lobby - Doors - Car - Panel Weight and other sensors Simulation equipment Computers	**Elevator System** - Lobby - Doors - Car - Panel Tools	**Elevator System** - Lobby - Doors - Car - Panel Design programs Computers ②	**Elevator System** - Lobby - Doors - Car - Panel Hammer Screwdriver Specialized tools Wiring	**Elevator System** - Lobby - Doors - Car - Panel Cleaner Property management offices/managers	**Elevator System** - Lobby - Doors - Car - Panel Drawing tools CAD Equipment

FIGURE 6.2

Multiple models of stakeholders' perspectives in an engineering design system; these stakeholders were concerned with the design of an elevator system. (Adapted from Benda, P. J., and Sanderson, P. M., 1998, Towards a dynamic model of adaptation to technological change, *Proceedings of the Australasian Computer Human Interaction Conference*, 244–251. © 1998 IEEE. With permission.)

	PATIENT			HOSPITAL		
	Private Life	Health	Cure	Cure	Care	Administration
Goals and Constraints	Working relations and conditions; Family relations; Goals and constraints of plans and commitments	Effects of illness and treatment on person's ability to meet subjective goals and criteria	Cure patient; Research, Training MDs; Public opinion; Legal, economic, and ethical constraints	Patient well being, physical and psychic care; Public opinion, economic and legal constraints	Laws and regulations of society, associations and unions; Workers' protection regulations etc.
Priority Measures, Flow of Values and Material	Personal economy, Probability of unemployment, cure, etc.	Probability of cure, priority measures, pace versus side-effects, etc.	Categories of diseases: Cost of treatments, patient suffering, research relevance	Flow of patients according to category; Treatment, and load on staff and facilities	Distribution of funds on activities; Flow of material and personnel to diseases, departments
General Functions and Activities	Work functions; Family relations; Living conditions	State of health; Diseases and possible treatments	Cure, diagnostics, surgery, medication, etc. Research, clinical, experiments	Board and lodging; Hygiene; Social Care, Physical support, transportation, etc.	Personnel and material administration, Accounting, sales and purchase
Physical Activities in Work, Physical Processes of Equipment	Physical work activities, spare time and sports activities; Homework; Transportation, etc.	Specific organic disorders and possible treatment, Previous illness and cures	Specific research and treatment procedures; Use of tools and equipment	Monitoring, treating, moving, cleaning and serving patients; Psychic Care	Processes in the administrative functions. Office and planning procedures
Appearance, Location and Configuration of Material Objects	Patient identification, age, address, profession, education, family members, etc.	Physical state of patient, weight, height, previous treatments, etc.	Material resources, patients, personnel, equipment; Medicine, tools, etc.	Facilities and equipment in patient quarters, kitchens, etc. Inventory of linen, food, etc.	Inventory of employees, patients, buildings, equipment, etc.

FIGURE 6.3

Work domains that must be considered in the treatment of patients in a hospital. (Adapted from Rasmussen, J., Pejtersen, A. M., and Goodstein, L. P., 1994, *Cognitive systems engineering*, Wiley, NY. Copyright © 1994 John Wiley & Sons. This material is reproduced with permission of John Wiley & Sons, Inc.)

column labels do not signify stakeholders, but the work domains that must be considered in the treatment of patients in a hospital independently of any stakeholders. These work domains include a patient's private life and health as well as a hospital's cure, care, and administrative fields of interest.

Figure 6.4 maps the stakeholders who are concerned with the treatment of patients in a hospital onto the various work domain models (Rasmussen et al., 1990). The stakeholders include patients' general medical practitioners as well as a hospital's medical doctors, general manager, head nurses, nurses, and assistant nurses. These stakeholders have different perspectives of the treatment of patients in a hospital. General medical practitioners are focused on planning the hospitalization of patients so that it fits in with patients' private lives, including their work, leisure, and social commitments, as well as their current states of health. A hospital's medical doctors are concerned with planning the hospitalization of patients so that it satisfies objectives related to curing patients, research, and training as well as legal, economic, and ethical constraints. Head nurses are focused on planning the hospitalization of patients so that it fits in with various laws and regulations such as those specifying acceptable working hours and conditions for hospital staff.

Three points are notable about Rasmussen et al.'s (1990, 1994) multiple models of stakeholders' perspectives in a health care system. First, a stakeholder may be concerned with more than one work domain. Nurses, for instance, are focused on the work domains of

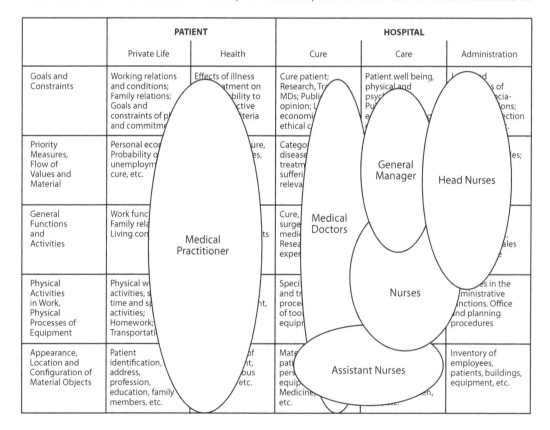

FIGURE 6.4
Stakeholders concerned with the treatment of patients in a hospital mapped onto the work domain models in Figure 6.3. (Adapted from Rasmussen, J., Pejtersen, A. M., and Schmidt, K., 1990, *Taxonomy for cognitive work analysis* (Risø-M-2871), Risø National Laboratory, Roskilde, Denmark. With permission.)

cure and care as well as administration (Figure 6.4). Second, several stakeholders may be concerned with the same work domain, although with different but overlapping aspects. For example, all of the stakeholders on the right of Figure 6.4 are involved in the work domain of care but have different, yet overlapping, spheres of concern. Third, stakeholders' object worlds may overlap at a variety of levels of abstraction, rather than primarily at lower levels as was the case for engineering design systems (Benda and Sanderson, 1998; Burns and Vicente, 2000). Figure 6.4 shows that a hospital's medical doctors and general manager have shared object worlds at higher and middle levels, medical doctors and nurses have shared object worlds at middle and lower levels, and medical doctors and assistant nurses have shared object worlds at lower levels.

Such insights are possible because Rasmussen et al.'s (1990, 1994) representation of the health care system differentiates between work domains and object worlds or, in other words, between work domains and stakeholders' views of those work domains. Although Burns and Vicente (1995, 2000) observed similar interactions between stakeholders, object worlds, and work domains in their study, their representation of the engineering design system (Figure 6.1) could not depict those interactions because their work domain models were not independent of stakeholders and their object worlds. Thus, the representation of the health care system is a more powerful model because it can accommodate observations of how work is achieved in that system.

The representation of the health care system is also consistent with the theoretical underpinnings of work domain analysis. Whereas work domains are event independent, stakeholders and their object worlds may vary as a function of the situation or over time. For example, the roles of head nurses and assistant nurses, and thus their object worlds, may shift or adapt in response to local contingencies, while the constraints associated with the work domains of cure, care, and administration in the health care system remain relatively constant. Hence, maintaining the distinction between work domains, which are event independent, and stakeholders and their object worlds, which are not, is important.

For these reasons, the guidelines for work domain analysis in Chapter 8 are formulated such that a representation consistent with that of the health care system, rather than either of the engineering design systems, is produced when modeling multiple stakeholders' perspectives. In brief, the solution lies in how the boundaries for a work domain analysis are conceptualized. That is, the boundaries must be defined such that they are independent of any specific actors, even when one wants to model the perspectives of particular stakeholders. This point is explained in more detail later in this section.

One other example of multiple models of stakeholders' perspectives is worth addressing. Torenvliet et al. (2008) conducted a study of a naval command-and-control system, specifically a frigate. Although the abstraction–decomposition space they produced appears to represent one of the object worlds relevant to the frigate, rather than the work domain of the frigate, this study is useful for elaborating on many of the preceding observations. The object world that Torenvliet et al. modeled is that of damage control (Figure 6.5). They also identified three other object worlds pertinent to the frigate, namely, combat systems engineering, marine systems engineering, and command and control. The stakeholders associated with these object worlds were not identified explicitly, but some of the relationships between the four object worlds were discussed.

One observation that Torenvliet et al. (2008) make is that the strongest overlap between the four object worlds occurs at the highest level of abstraction. That is, the functional purposes of the object worlds are highly similar, although not the same. On the other hand, the weakest overlap occurs at the lowest two levels. In fact, the physical functions and physical forms of the object worlds are unique.

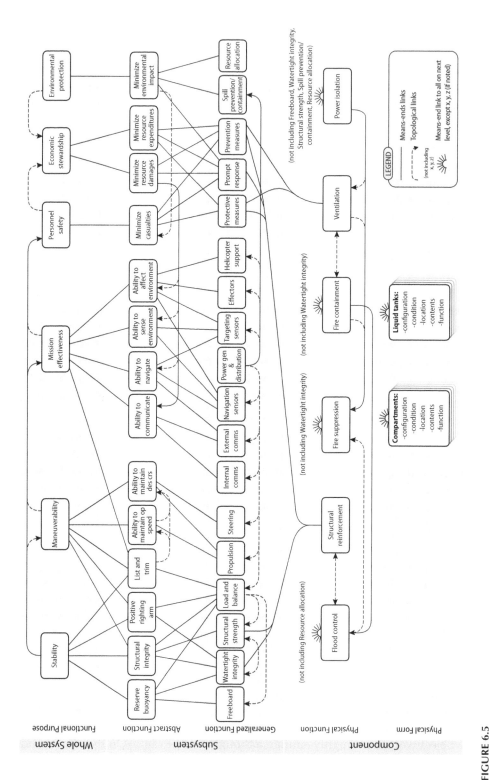

FIGURE 6.5

Abstraction–decomposition space of damage control, one of the object worlds of a naval command-and-control system. (From Torenvliet, G. L., Jamieson, G. A., and Chow, R., 2008, Object worlds in work domain analysis: A model of naval damage control, *IEEE Transactions on Systems, Man, and Cybernetics—Part A: Systems and Humans*, 38(5), 1030–1040. © 2008 IEEE. With permission.)

These observations are different from those made in relation to engineering design systems (Benda and Sanderson, 1998; Burns and Vicente, 2000). In Burns and Vicente's (2000) study, the strongest overlap between the object worlds was at lower levels of abstraction, and the weakest overlap was at higher levels. Torenvliet et al.'s (2008) results, though, are not inconsistent with Rasmussen et al.'s (1990, 1994) findings that the object worlds of some of the stakeholders in the health care system overlapped mainly at higher rather than lower levels.

Torenvliet et al. (2008) suggest that the nature of the overlap between stakeholders' object worlds may depend on the dominant form of work coordination in a system. When coordination is primarily bottom-up, or constrained by a system's physical objects, the strongest overlap may occur at lower levels of abstraction. Engineering design systems are an example of this case. In contrast, when coordination is principally top-down, or constrained by a system's objectives, the strongest overlap may occur at higher levels. The naval command-and-control system falls within this category.

It is worth adding that Rasmussen et al.'s (1990, 1994) study of a health care system demonstrates that both bottom-up and top-down coordination may be present within the same system. The coordination between a hospital's medical doctors and assistant nurses, which is organized around physical entities such as patients and medical equipment, may be described as bottom-up. Conversely, the coordination between a hospital's medical doctors and general manager, which is organized around objectives such as satisfying economic and legal constraints, may be described as top-down.

Torenvliet et al. (2008) utilize the concept of object worlds to explain some of the differences between their model of a frigate and two other models of frigates, which were produced by Linegang and Lintern (2003) and Burns et al. (2005). If all three models are considered to be of the same system, it is reasonable to ask why there are differences in the content of the models. According to Torenvliet et al., the three models are different because they focus on different object worlds. Their own model focuses on damage control (Figure 6.5), Burns et al.'s model focuses on command and control,[*] and Linegang and Lintern's model focuses on antisubmarine warfare as well as damage control (Figure 6.6). Neither Burns et al. nor Linegang and Lintern invoke the concept of object worlds in describing their model. Nevertheless, Torenvliet et al. conclude that the selection—either implicitly or explicitly—of object worlds for the three analyses is the primary explanation for the differences in the content of the models. They also acknowledge that some of the differences may be due to dissimilarities in project objectives. While the purpose of their own model was to assist with simulation development, the other models were aimed at supporting overall system design (Burns et al., 2005) and display design (Linegang and Lintern, 2003).

Although it may not be recognized or acknowledged explicitly, all work domain models represent the perspective of particular actors at some level. Without choosing a viewpoint to define the boundaries of an analysis, it would be difficult to produce a model at all, let alone one that is suitable for the intended application. The home in Figure 2.7 is modeled from the perspective of actors living in the home. However, a home could also be modeled from the viewpoint of actors building the home. The two models would be very different. One would describe the constraints associated with living in the home, whereas the other would describe the constraints associated with building the home. Which model is produced would depend on the purpose of the analysis.

---

[*] See Figure 6.9 for Burns et al.'s model of a frigate.

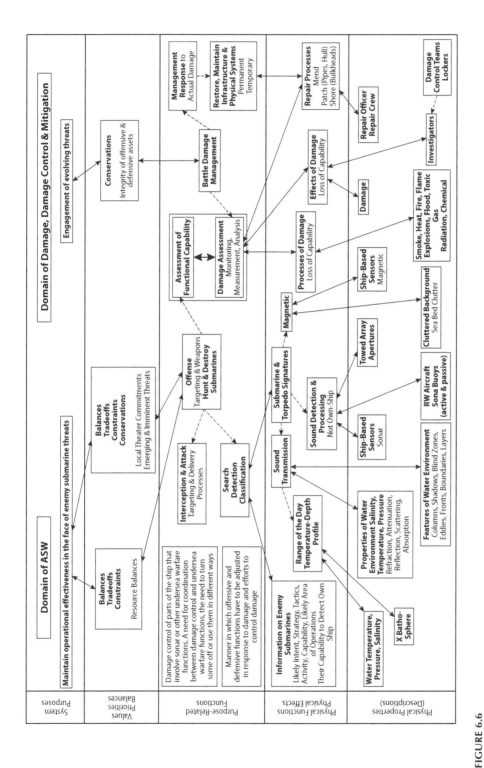

FIGURE 6.6

Partial abstraction–decomposition space of antisubmarine warfare and damage control in a naval command-and-control system. Arrows with broken lines point to nodes at lower levels of decomposition relative to the nodes from which they originate. (Adapted from Linegang, M. P., and Lintern, G., 2003, Multifunction displays for optimum manning: towards functional integration and cross-functional awareness, *Proceedings of the Human Factors and Ergonomics Society 47th Annual Meeting,* 1923–1927. Copyright © 2003 by the Human Factors and Ergonomics Society. All rights reserved.)

It is important, though, that the actors defining the viewpoint of an analysis are conceptualized in such a way that they are event independent. That is, the actors must be relevant regardless of the circumstances. One way of thinking about this is that the actors must be generic or nonspecific. For instance, the actors building a home could be inhabitants, qualified builders, or robots; 'actors building a home' is a generic category, whereas inhabitants, qualified builders, or robots are specific actors. Depending on the situation, any of these specific actors could be building the home, but, irrespective of which of these actors is involved, the constraints associated with building a home remain relatively constant. If an analysis was based on a specific actor's perspective, the result would not be a work domain model but a model of a particular stakeholder's object world.

Additional examples of multiple models of stakeholders' perspectives can be found in the literature. These include models of other engineering design systems (Pejtersen et al., 1997; Rasmussen, 1988) as well as those of a library (Rasmussen et al., 1994), a medical system (Benda and Sanderson, 1999), and a gaming system (Burns and Proulx, 2002).

Multiple Problem Facets

Another reason for building multiple models of the same system is to emphasize the distinct facets of a problem with which actors are required to deal. Such representations also highlight the fundamental differences in control that actors have over various elements of their work domain. These representations allow designs that recognize the distinctions in the nature of a problem or degree of control to be devised systematically.

Generally, analysts have constructed multiple models of problem facets for "loosely bound" (Burns et al., 2005, p. 603) rather than tightly bound systems. Rasmussen et al. (1987) explain the distinction between these two classes of systems by comparing process control (e.g., nuclear power plants and petrochemical refineries) with emergency management systems (e.g., firefighting and disaster relief organizations). A process control system has relatively well-defined boundaries because its actors have full control over all of the entities or processes they are responsible for managing (e.g., plant components and processes). In addition, the disturbances they have to handle originate from internal faults (e.g., chemical spills). This type of system may be described as tightly bound. In contrast, an emergency management system may be described as loosely bound. Its actors do not have full control over all of the elements for which they are accountable, including both the sources of accidents (e.g., industrial process plants) and the targets of damage (e.g., people). Also, these actors are responsible for dealing with disturbances that originate externally (e.g., fires, floods, radiation release).

Rasmussen et al. (1987) observe that whereas process control systems can be represented faithfully with a single model, emergency management systems require multiple models. Figure 6.7 provides an overview of Rasmussen et al.'s representation of an emergency management system for nonnuclear industries that shows that it consists of two models.[*] One model, called the domain of potential risk, includes the potential sources of disturbances (e.g., industrial plants, forests, rivers) and the potential targets of damage (e.g., people, cities). The other model, called the domain of emergency management resources,

---

[*] See Figures 6.13 and 6.14 for the two models of the nonnuclear emergency management system.

Domain of Potential Risk	Domain of Emergency Management Resources
Goals, values, and constraints; Laws, regulations and public opinion	
Criteria for setting priority; Risk, economic and social effects	Criteria for setting priority; Flow of monetary values, and manpower
General accident categories Fire, flooding, explosions, etc.	General resources Medical care, evacuation, firefighting
Physical processes in potential accidents as specified by analysis or prior events	Physical functioning of tools and equipment; Interaction with target
Topographical, demographical characteristics of potential danger sources and localities	Appearance, location, and configuration of material, tools and resources

FIGURE 6.7

Overview of multiple models of the distinct facets of the problem of nonnuclear emergency management. The full models are shown in Figures 6.13 and 6.14. (Adapted from Rasmussen, J., Pedersen, O. M., and Grønberg, C. D., 1987, *Evaluation of the use of advanced information technology (expert systems) for data base system development and emergency management in non-nuclear industries* (Risø-M-2639), Risø National Laboratory, Roskilde, Denmark. With permission.)

includes the resources for counteracting or mitigating emergencies (e.g., fire brigades, transport vehicles, hospitals).

Rasmussen et al.'s (1987) representation differentiates potential *emergencies* from *resources* for managing those emergencies. As a result, it partitions the problem of nonnuclear emergency management into two distinct facets. One facet relates to identifying, assessing, and prioritizing emergencies (domain of potential risk). The other relates to identifying, assessing, and prioritizing resources for managing those emergencies (domain of emergency management resources). According to Rasmussen et al. (1987), the top level of abstraction is shared across the two models because policy about the acceptable level of risk to society is formulated at this level, which requires consideration of both the nature of the potential risks in the environment and the nature of the resources required for managing those risks.

The representation in Figure 6.7 also reflects the differences in control that the emergency management system has over the entities or processes it has the task of managing, although Rasmussen et al. (1987) do not make this point explicitly. Specifically, the representation differentiates elements over which the emergency management system has little or no control (domain of potential risk) from those over which it has greater control (domain of emergency management resources).

Following Rasmussen et al. (1987), Hajdukiewicz et al. (1999) developed multiple models of problem facets for a military command-and-control system, specifically an air operations center. Figure 6.8 shows that this representation consists of two models. The model labeled the domain of military encounter includes the potential risks or threats posed by enemy forces, whereas the model labeled the domain of friendly resources includes military assets for combating those risks or threats.

	Domain of Military Encounter	Domain of Friendly Resources
Functional Purpose	Goal: To accomplish a military course of action deemed to be in the best interest of the country by the national command authority	
Abstract Function	Priority criteria and given level of acceptable risk balance (e.g., public opinion, congressional support, international relations, rules and regulations)	Time, resource balance within engagement region, probability of successful engagement
Generalized Function	Urgency of military threats, situation assessment; information coordination; decision coordination and collaboration	Transportation, effectiveness of military resources for engagement
Physical Function	Functional capabilities and limitations of military threats, trust/belief in information, risk/outcome estimation and projection; information assurance	Combatant skills, functional capabilities and limitations of equipment and weapons
Physical Form	Location of military threats, Information location, method, and form of delivery; communication modes; collaboration modes	Location of people, equipment, weapons, geographic landscape, air space landscape, weather

FIGURE 6.8

Multiple models of the distinct aspects of the problem of military command and control in an air operations center. (Adapted from Hajdukiewicz, J. R., Burns, C. M., Vicente, K. J., and Eggleston, R. G., 1999, Work domain analysis for intentional systems, *Proceedings of the Human Factors and Ergonomics Society 43rd Annual Meeting*, 333–337. Copyright © 1999 by the Human Factors and Ergonomics Society. All rights reserved.)

Like Rasmussen et al. (1987), Hajdukiewicz et al. (1999) distinguish between potential risks (or emergencies) and resources. Thus, the problem of military command and control in an air operations center is portrayed as having two distinct aspects, which are similar to those identified for nonnuclear emergency management. One aspect involves identifying, assessing, and prioritizing the risks posed by enemy forces (domain of military encounter) and the other involves identifying, assessing, and prioritizing the military assets for managing those risks (domain of friendly resources). As in the representation of nonnuclear emergency management (Figure 6.7), the first level of abstraction is shared across the two models of the military command-and-control system, although Hajdukiewicz et al. do not discuss why.

Hajdukiewicz et al.'s (1999) representation also indicates the fundamental distinctions in control that the air operations center has over the various entities or processes it is responsible for managing, paralleling the depiction of the nonnuclear emergency management system (Figure 6.7). That is, the representation distinguishes between elements over which the air operations center has limited or no control (domain of military encounter) and those over which it has more control (domain of friendly resources). However, like Rasmussen et al. (1987), Hajdukiewicz et al. do not address this point explicitly.

Subsequent to the work of Hajdukiewicz et al. (1999), Burns et al. (2005) also constructed multiple models of problem facets for a military command-and-control system, in this case consisting of a frigate instead of an air operations center. In contrast to Hajdukiewicz et al.'s representation (Figure 6.8), which comprises two models, Burns et al.'s representation has three models. The first depicts the frigate (Figure 6.9), the second describes the natural environment (Figure 6.10), and the third represents contacts or entities external to the frigate with friendly, hostile, or neutral intent (Figure 6.11).

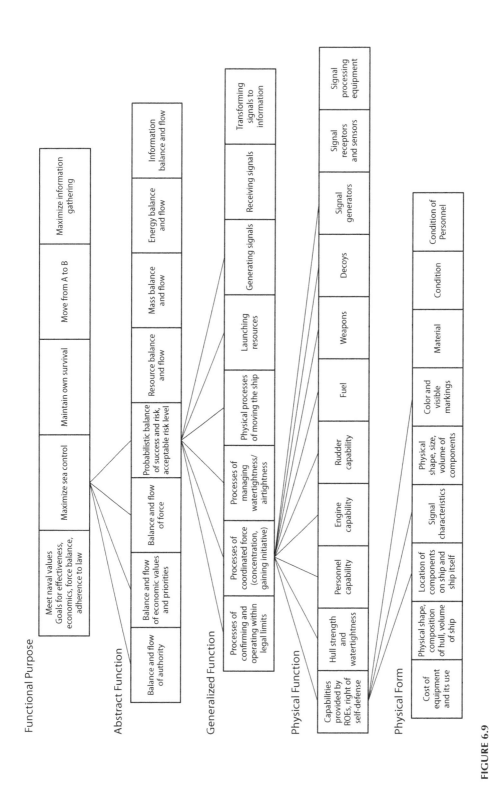

FIGURE 6.9
Multiple models of distinct aspects of the problem of military command and control on a frigate: model of the frigate. (Adapted from Burns, C. M., Bryant, D. J., and Chalmers, B. A., 2005, Boundary, purpose, and values in work-domain models: Models of naval command and control, *IEEE Transactions on Systems, Man, and Cybernetics—Part A: Systems and Humans*, 35(5), 603–616. © 2005 IEEE. With permission.)

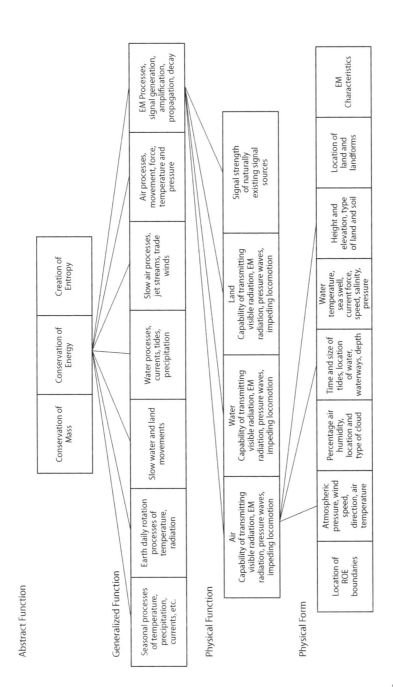

FIGURE 6.10
Multiple models of distinct aspects of the problem of military command and control on a frigate: model of the natural environment. (From Burns, C. M, Bryant, D. J., and Chalmers, B. A., 2005, Boundary, purpose, and values in work-domain models: Models of naval command and control, *IEEE Transactions on Systems, Man, and Cybernetics—Part A: Systems and Humans*, 35(5), 603–616. © 2005 IEEE. With permission.)

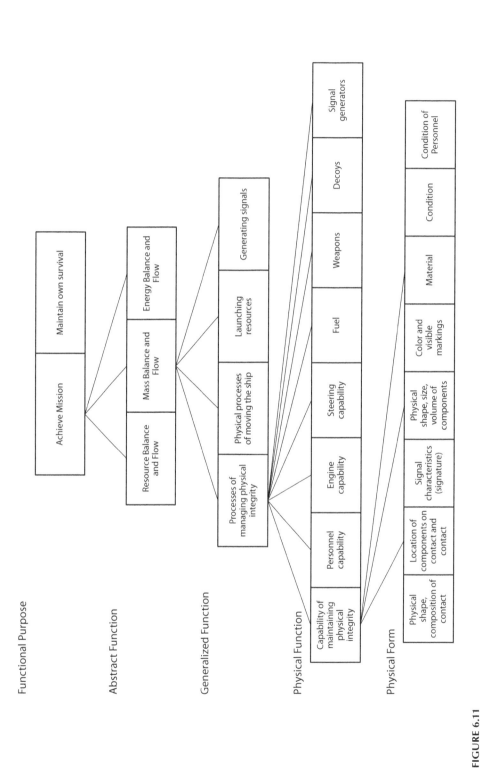

FIGURE 6.11
Multiple models of distinct aspects of the problem of military command and control on a frigate: model of contacts. (From Burns, C. M., Bryant, D. J., and Chalmers, B. A., 2005, Boundary, purpose, and values in work-domain models: Models of naval command and control, *IEEE Transactions on Systems, Man, and Cybernetics—Part A: Systems and Humans*, 35(5), 603–616. © Copyright 2005 IEEE. With permission.)

Although Burns et al.'s (2005) representation contains three models, it portrays the problem of military command and control in a frigate as having two distinct facets, which are similar to those for nonnuclear emergency management (Rasmussen et al., 1987; Figure 6.7) and military command and control in an air operations center (Hajdukiewicz et al., 1999; Figure 6.8). One facet, represented by the models of the natural environment and contacts, concerns situation and threat assessment (Burns, Bisantz, and Roth, 2004). The other facet, represented by the model of the frigate, concerns resource management.

According to Burns et al. (2005), one reason for developing separate models of the frigate, the natural environment, and the contacts was to highlight the basic differences in control that the frigate has over its own systems compared with the two other elements of the work domain. Whereas the frigate has complete control over its own systems, it has no control over the natural environment, and it does not always have control over the contacts. Another reason was to emphasize that the frigate, the natural environment, and the contacts do not have shared purposes. A contact can be in conflict with the frigate and attempt to exert control over it by damaging it with weapons. Furthermore, the natural environment can constrain the movement of the frigate and its use of sensors and weapons. Therefore, the purposes level of abstraction is not shared across the models in the representation of the naval command-and-control system, unlike in the representations of nonnuclear emergency management (Rasmussen et al., 1987; Figure 6.7) and military command and control in an air operations center (Hajdukiewicz et al., 1999; Figure 6.8).

Finally, it is important to mention that analysts do not always develop multiple models of loosely bound systems. Bisantz et al.'s (2003) representation of a naval destroyer, for instance, comprises a single model (Figure 6.12). Because of the similarities between this naval destroyer and the frigate modeled by Burns et al. (2005), the representations of the two systems were compared by Burns et al. (2004). This comparison revealed some of the benefits and limitations of composing multiple models of a system to emphasize the distinct aspects of a problem versus creating a single, integrated model.

Burns et al. (2004) observed that the main benefit of differentiating the naval vessel from the natural environment and contacts, as was done for the frigate (Burns et al., 2005; Figures 6.9, 6.10, and 6.11), was that it portrayed the distinctions in control that the frigate has over its own systems compared with the two other elements of the work domain. In addition, it allowed more detailed representations to be produced of each element. Conversely, constructing a single, integrated model of the naval vessel, natural environment, and contacts, as was done for the destroyer (Bisantz et al., 2003; Figure 6.12), made the interactions between these elements more apparent. For example, the model of the destroyer highlighted interactions between the performance of sensor systems on the naval vessel and the environmental conditions.

To summarize, analysts may build multiple models of loosely bound systems to emphasize the distinct facets of a problem or the fundamental differences in control that a system has over elements of the work domain. This makes it possible to generate designs that respect these key distinctions. On the other hand, a single, integrated model may be beneficial for emphasizing interactions between the various aspects of a problem or elements of a work domain. As a result, designs may be devised that recognize these important interactions. Chapter 8, which presents guidelines for performing work domain analysis, illustrates how choices may be made between these two options in relation to a home. The case studies of military systems in Chapters 9, 10, and 11 demonstrate how such decisions may be made in light of a variety of applications in industrial settings.

Some other examples of multiple models of problem facets include representations of an emergency management system in the nuclear industry (Moray et al., 1992), an emergency

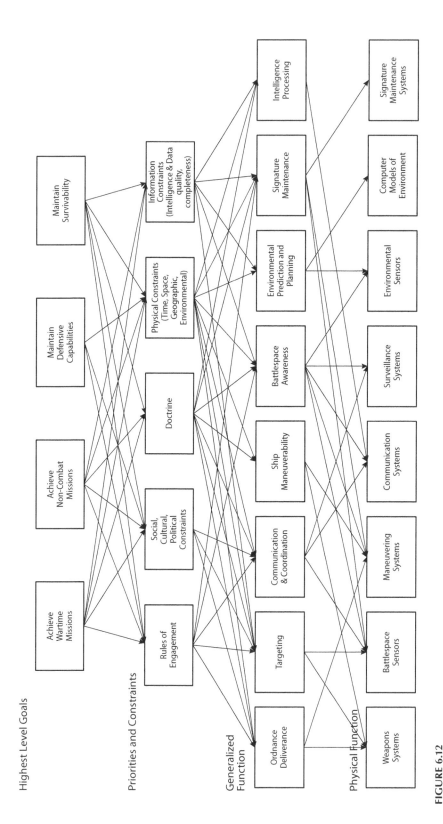

FIGURE 6.12
Single, integrated model of a naval destroyer. (Adapted from Bisantz, A. M., Roth, E., Brickman, B., Gosbee, L. L., Hettinger, L., and McKinney, J., 2003, Integrating cognitive analyses in a large-scale system design process, *International Journal of Human-Computer Studies, 58*, 177–206. Copyright 2003. With permission from Elsevier.)

ambulance dispatch management system[*] (Hajdukiewicz et al., 1999), and a military command-and-control system for employing uninhabited aerial vehicles in suppression of enemy air defense missions[†] (Rasmussen, 1998).

Multiple Models and Decomposition

Two brief observations are worth making about the relationship between multiple models, whether they be of stakeholders' perspectives or problem facets, and decomposition. One is that modeling a system at different levels of detail, or decomposition, may be viewed as composing multiple models of a system. In this book, however, the term *multiple models* is reserved for referring to representations of stakeholders' perspectives or problem facets.

The other is that the examples presented above show that analysts typically construct a series of abstraction hierarchies when developing multiple models. Nevertheless, a series of abstraction–decomposition spaces may be constructed instead. Figures 6.13 and 6.14 show a pair of abstraction–decomposition spaces that Rasmussen et al. (1987) used to model the distinct aspects of the problem of nonnuclear emergency management that were addressed previously (Figure 6.7). The model in Figure 6.13 represents the domain of potential risk, while that in Figure 6.14 represents the domain of emergency management resources.

Summary

In most cases, a single work domain model is created of a system, but there are times when it may be beneficial to develop multiple models instead. One reason it may be useful to do this is to capture multiple stakeholders' perspectives of a problem. This makes it possible to conceive of designs for one stakeholder that do not have undesirable consequences for other stakeholders, and it enables designs to be invented that support better coordination among stakeholders. In building this type of representation, it is essential that the distinction between work domains, which are event independent, and stakeholders and their object worlds, which are not, is respected. This requirement is consistent with the theoretical underpinnings of work domain analysis. It also allows analysts to portray the following: that a stakeholder may be concerned with more than one work domain; that several stakeholders may be focused on the same work domain, although with different but overlapping aspects; and that stakeholders' object worlds may overlap at a variety of levels of abstraction, perhaps depending on the form of coordination in a system. The requirement to respect this distinction has implications for the way in which the boundaries of a work domain analysis are conceptualized. In essence, the boundaries must be formulated such that they are independent of specific actors, even when the aim is to capture

---

[*] See Figure 5.6a for multiple models of problem facets for the emergency ambulance dispatch management system.

[†] See Figure 4.10 for the abstraction–decomposition space of the resources domain for the military command-and-control system for suppression of enemy air defense missions. The model of the threat (or risks) domain is not presented in this book.

	National Overview and Patterns	Emergency Classes	Companies and Installations	Specific Production Plants and Equipment	Processes, Substances and Components
Goals and Constraints		Risk pattern in terms of social and economic consequences with reference to features of established policies and public opinion			
Priority Criteria Economy, Risk, Man Power Flow	National pattern, geography and demography	Risk pattern as related to industrial branches	Risk pattern of individual installations and plants	Risk related to specific processes	Risk related to specific materials, substances and components
		Risk measures in terms of economy, probability and other abstract measures suitable for setting priorities			
General Functions and Operations		Accident potential in general terms; fire, explosion, flooding, toxication			
	Relation to geographical regions or population features	Relation to industrial activities or to population groups	Relation to specific process plants or installations	Functional and accidental mechanisms of specific processes	Risk classes related to categories of processes, substances, and material
Function of Specific Installations, Groups and Equipment		Physical processes and mechanisms behind accidents, causation, propagation, potential for interaction with accident control measures			
	National and geographical patterns, meteorological data, water streams, other propagation characteristics	General data on industrial practices, processes and accidental mechanisms. Safety measures	Functional information on specific plants, accident potential and mechanisms, safety measures	Relation to specific manufacturing processes	Properties of substances and materials
Material Locations Configurations Appearance		Locations, topography, physical design and appearance			
	National pattern of potential sources and population, propagation routes, road and barrier topography	Distribution according to branches and risk categories	Location of specific plants and installations. Drawings of buildings and access routes, maps of likely propagation paths	Location of specific process equipment, identification data, transport and access information	Information for identification and location of material, substances, and components. Personal data

FIGURE 6.13

Multiple models of the distinct facets of the problem of nonnuclear emergency management: abstraction–decomposition space for the domain of potential risk. (Adapted from Rasmussen, J., Pedersen, O. M., and Grønberg, C. D., 1987, *Evaluation of the use of advanced information technology (expert systems) for data base system development and emergency management in non-nuclear industries* (Risø-M-2639), Risø National Laboratory, Roskilde, Denmark. With permission.)

the perspectives of particular stakeholders. Otherwise, the resulting representation would not be a series of work domain models; it would be a depiction of particular stakeholders' object worlds.

Multiple models of the same system may also be worth developing to emphasize the distinct facets of a problem that actors are required to deal with. This category of multiple models is normally applicable for loosely bound rather than tightly bound systems. By formulating separate models of risks and resources for managing those risks, for instance, problems in loosely bound systems may be revealed as having two distinct aspects. These are risk (or situation) assessment and risk (or resource) management. Such representations may also highlight the fundamental differences in control that actors have over various

	National Overview and Patterns	Activity Categories Emergency Classes	Organizations and Institutions	Emergency Task Forces	Individual People and Major Tools
Goals and Constraints	National laws and government agency regulations	Goals and constraints for measures against: fires, floods, traffic accidents, etc.	Goals and targets for services and institutions: hospitals, fire brigades, "Falck"	Goals and targets for groups and task forces	Exposure limits for individuals, regulation data
Priority Criteria Economy, Risk, Man Power Flow		Criteria and measures for priority setting Flow, accumulation, and turn-over of funding, man power, and material according to: - Risk categories	- Services	- Task forces	- Individuals and equipment
General Functions and Operations		Available resources for general emergency control functions: Fire fighting, medical care, transportation and evacuation, etc. General overview of resources. General rules and heuristics for counter measures	Resources specified with reference to organizations, institutions. General institutional rules and practices	Resources of identified task forces, groups, and operational units and institutions	Capabilities of equipped individuals and major tools
Function of Specific Installations, Groups and Equipment		Physical functioning, capabilities, and limitations of emergency control mechanisms		Physical functions and capabilities of tools as available to task forces and groups. Instructions and procedures, standing orders	Physical Characteristics and limitations of tools. Information on possible, unacceptable interaction with media and installations (chemical, electrical, etc.). Procedures and practices
Material Locations Configurations Appearance	Road system with data on traffic and load capacity	Locations, descriptions, identification of items, forces, groups Geographical location of services and institutions, access routes	Drawings of premises of individual institutions. Drawings of buildings. Inventory lists of service stations	Inventory, locations, identifying characteristics of equipment, tools, and members of task forces	Drawings of equipment, with size and weight data

FIGURE 6.14

Multiple models of the distinct facets of the problem of nonnuclear emergency management: abstraction–decomposition space for the domain of emergency management resources. (Adapted from Rasmussen, J., Pedersen, O. M., and Grønberg, C. D., 1987, *Evaluation of the use of advanced information technology (expert systems) for data base system development and emergency management in non-nuclear industries* (Risø-M-2639), Risø National Laboratory, Roskilde, Denmark. With permission.)

elements of the work domain. Designs that recognize these basic distinctions in the nature of a problem or degree of control may then be devised systematically. Nevertheless, analysts do not always build multiple models of loosely bound systems. A single, integrated model has the advantage of exposing interactions between the distinct aspects of a problem or elements of a work domain, which may also be beneficial for design.

Finally, it is worth remembering that the term *multiple models* is reserved here to describe representations of multiple stakeholders' perspectives or problem facets. It is not used to refer to models of a system at different levels of decomposition. In addition, it is possible to use a series of abstraction–decomposition spaces, rather than abstraction hierarchies, to assemble multiple models.

7

Activity: Whether or Not to Model?

OVERVIEW   This chapter addresses four key questions pertaining to the relationship between the analysis of activity and work domain analysis: whether activity can be included in a work domain model; whether nouns or verbs should be used for representing constraints in this type of model; whether control systems may be incorporated into a work domain model; and whether there is an overlap between work domain analysis and control task analysis, the second dimension of cognitive work analysis, which focuses on the modeling of activity. These matters, which were not resolved in earlier texts, are important to settle because they have implications for the scope and accuracy of a work domain model. This, in turn, can affect how well designs based on the model support workers in adapting to the demands of a broad range of situations, including those that are unforeseen.

Can Activity Be Included in a Work Domain Model?

The question of whether activity can be incorporated into a work domain model is fundamental to work domain analysis, but the texts by Rasmussen et al. (1994) and Vicente (1999) may appear to offer different perspectives on this matter. Through various statements, Rasmussen et al. imply that activity can be represented in this kind of model. For instance, they mention that work domain analysis produces an "inventory of objectives, functions, activities, and resources" (p. 35), and they label one of the five levels of abstraction "Physical Processes and Activities" (p. 38). They also state that the general functions level "comprises the properties necessary and sufficient to identify the functions and activities to be coordinated" (p. 39).

Vicente (1999), on the other hand, argues that a work domain model should not represent actions and action means–ends relations but, instead, represent the objects of action and structural means–ends relations. As an example, he recounts that this type of model should contain objects, such as 'furnace' and 'fireplace,' and structural means–ends relations, such as "Furnace and fireplace are both *objects* that can be used to achieve warmth" (p. 162; italics in the original). However, it should not include actions, such as 'going down to the basement' and 'lighting the fireplace,' and action means–ends relations, such as "Going down to the basement and then lighting the fireplace are both means for achieving warmth" (p. 162). Furthermore, in the context of Simon's (1981) parable about an ant on the beach,[*] Vicente points out that work domain analysis models the properties of the beach, and that activity is not a property of the beach, but a property of the ant.

These types of statements by Rasmussen et al. (1994) and Vicente (1999) may suggest that they hold conflicting views about the modeling of activity in work domain analysis. However, I do not believe this is the case. Like Vicente, Rasmussen et al. do not include actions or action means–ends relations in their work domain models. In other words, their

---

[*] Chapter 2 describes the parable of an ant on the beach.

Purposes and Constraints	Prosperity, Well-being
Abstract Functions	Implications of the functions in terms of values and resources absorbed
General Functions	Functions of a city: transport, trade, health care, administration, public education
Physical Processes and Activities	Processes of a city: moving goods and people, sleeping, feeding, shaping and assembling products, chemical and physical production processes
Physical Form and Configuration	Material objects: people and houses, furniture and tools, cars and street lamps

FIGURE 7.1

Abstraction hierarchy of a city. (Adapted from Rasmussen, J., Pejtersen, A. M., and Goodstein, L. P., 1994, *Cognitive systems engineering*, Wiley, New York. Copyright © 1994 John Wiley & Sons. This material is reproduced with permission of John Wiley & Sons, Inc.)

models do not contain actions or sequences of action for achieving various goals. Instead, their comments reflect the fact that, among other kinds of constraints, they incorporate the processes or functions *afforded* by a system into their models. For example, in the context of Simon's (1981) parable about an ant on the beach, Rasmussen et al. might list 'walking' and 'sleeping' as processes or functions afforded by the beach in a model. Their intention in including such concepts would not be to represent the activity of the ant, but to represent the affordances of the beach. Hence, like Vicente, Rasmussen et al. portray the objects of action (or the structural properties of the environment) and structural means–ends relations in their models.

Figure 7.1 shows an abstraction hierarchy of a city from the text by Rasmussen et al. (1994). The label of the fourth level is 'physical processes and activities,' and the third level is called 'general functions.' Regardless, this model does not represent actions or action means–ends relations for achieving the city's goals, which could be to increase foreign investment or improve facilities for aged care. Instead, as well as other types of constraints, it describes the processes or functions afforded by various physical objects to fulfill the city's purposes, which include prosperity and well-being.* Thus, the model of the city represents objects of action and structural means–ends relations.

The fact that actions and action means–ends relations are not depicted in a work domain model should not be taken to mean that activity is inconsequential within the framework of cognitive work analysis. On the contrary, models of activity are regarded as essential for many design problems (Rasmussen, 1986; Rasmussen et al., 1994; Vicente, 1999). However, activity is not the focus of work domain analysis but the focus of control task analysis and strategies analysis, the second and third dimensions of the framework. These dimensions are considered further in the Appendix.

---

* As explained in Chapter 3, goals are dynamic and may vary as a function of the situation, whereas purposes are more stable over time.

As discussed in Chapter 2, the aim of work domain analysis is to develop a model of the functional structure of the physical, social, or cultural environment of actors. The structural properties of the environment place constraints on actors' behavior, but within these constraints actors have many possibilities for action. As these properties remain relatively constant, unlike actors' behavior, a work domain model is event independent. That is, it identifies the constraints on actors, as well as the possibilities for action open to them, over a wide variety of situations. With such a model, it becomes possible to produce designs that support actors in dealing with a broad range of events, including those that are novel or unanticipated.

Incorporating actions and action means–ends relations into a work domain model is undesirable because representations of activity are event dependent (Vicente, 1999). The actions or action sequences necessary in any particular situation depend on the goals that are relevant. This means that models of activity can only be developed when the goals are well defined and, therefore, for situations that are known or can be anticipated. The actions or action sequences required in novel or unanticipated situations cannot be specified because the goals in these situations cannot be predicted a priori. Consequently, models of activity cannot be used to create designs that support actors in handling unforeseen events, which pose the most significant threats to a system's effectiveness.[*]

Should Nouns or Verbs Be Used in a Work Domain Model?

Another issue that requires clarification is whether nouns or verbs should be used to represent constraints in a work domain model. As before, Rasmussen et al. (1994) and Vicente (1999) may appear to have different views on this topic. Rasmussen et al. often use verbs in their work domain models. For example, their abstraction hierarchy of a city (shown in Figure 7.1) includes verbs such as "shaping" and "assembling" (p. 43), and their model of a health care system[†] contains verbs such as "monitoring" and "treating" (p. 47). This is another reason why they may be misinterpreted as portraying activity in their models.

Vicente (1999), in contrast, encourages analysts to use nouns in their work domain models. He argues that nouns are more appropriate for describing objects of action, whereas verbs are more appropriate for describing actions. Hence, nouns are better suited for modeling work domains, while verbs are better suited for modeling activity.

To emphasize the difference between using nouns and using verbs in a work domain model, consider the following examples from the abstraction–decomposition space of a home (Figure 2.7). One of the functional purposes in this model is *environmental protection*. The word *protection* in this entry is a noun. The use of the verb form to represent this constraint would result in the description 'protect the environment.' Similarly, one of the purpose-related functions in this model is *preparation of meals and beverages*, where the term *preparation* is a noun. Adopting the verb form for this entry instead would lead to the description 'prepare meals and beverages.'

It is certainly more accurate to label constraints that represent objects of action with nouns rather than verbs. (The constraints in a work domain model describe the structural properties of actors' environments, not their behavior.) Moreover, the practice of using

---

[*] As discussed in Chapter 1, unforeseen events place significant pressure on a system's effectiveness.
[†] See Figure 6.3 for the model of the health care system.

nouns as often as possible will help analysts ensure that they are modeling objects of action and structural means–ends relations. As mentioned earlier, the objective of work domain analysis is to develop an event-independent representation of the constraints on actors' behavior. Such a model can be used to formulate designs that help actors in managing the demands of a wide range of situations, including those that are unpredictable. By using nouns whenever possible, analysts are less likely to incorporate actions and action means–ends relations inadvertently into their models.

There are, however, two difficulties with always using nouns in a work domain model. One problem is that some constraints may be more naturally described with verbs rather than nouns. In relation to the model of a home (Figure 2.7), for example, it may seem more natural to say that an objective of this system is to 'protect the environment' rather than *environmental protection*. In this case, the verb form is not intended to signify activity but an objective, which is a relatively permanent property of a system that places constraints on actors. Constraints that are labeled with nouns when they are more naturally described with verbs may result in representations that seem contrived or artificial to analysts as well as to subject matter experts. Furthermore, when verbs are more natural labels for constraints, it can be difficult or time consuming for analysts to identify suitable nouns for describing those constraints.

Another problem is that many words in the English language can be either nouns or verbs. *Exercise* and *rest* are two such examples in the home model (Figure 2.7). Normally, the grammatical form that these words assume depends on their function within a sentence. However, given that constraints are generally represented with single words or short phrases, this context is not provided in most work domain models. The alternative, then, is to find labels for constraints that are nouns irrespective of context. This option, though, can also be difficult or time consuming for analysts.

It is important to recognize that the issue of whether to use nouns or verbs in a work domain model only arises because of the practice of drawing on single words or short phrases to represent constraints rather than entire sentences or paragraphs. This practice is partially due to the space limitations of the four main formats for presenting a work domain model, that is, the graphical abstraction–decomposition space, the graphical abstraction hierarchy, the tabular abstraction–decomposition space, and the tabular abstraction hierarchy.* Another reason for employing single words or short phrases in a model is that it results in manageable representations of highly complex systems, which are conducive to design. Nevertheless, constraints are often described more meaningfully with whole sentences or paragraphs. Therefore, irrespective of whether analysts use nouns or verbs to label the constraints in a model, they may find it beneficial to develop a glossary that describes those constraints in more detail. For this reason, the guidelines in Chapter 8 suggest the use of a glossary to supplement a work domain model.

In summary, the practice of using nouns whenever possible to label constraints will help analysts ensure that their work domain models represent objects of action and structural means–ends relations, instead of actions and action means–ends relations. However, if labeling certain constraints with nouns results in descriptions that seem contrived or artificial, or if finding suitable nouns to label particular constraints is challenging, analysts should not rule out the use of verbs, although they need to be careful that they do not inadvertently incorporate activity into their models. Analysts may find it helpful to supplement their models with a glossary that describes the constraints in more depth, irrespective of whether they are labeled with nouns or verbs.

---

* See Figure 2.18 for the four main formats for presenting a work domain model.

Can Control Systems Be Included in a Work Domain Model?

An issue that is the focus of some debate is whether control systems can be represented in a work domain model. For the purposes of this discussion, a control system may be viewed simply as a component of a system that manages (e.g., directs, regulates) the behavior or operation of other components. Control systems can be automated or manual. An example of an automated control system is a thermostat that senses the temperature of a room and adjusts a heater to keep the temperature at a certain level. An example of a manual control system, on the other hand, is a person who moves or manipulates a paper fan to direct currents of air toward an object. A person who manages the behavior of other people may also be considered a manual control system.

The debate about whether control systems can be incorporated into a work domain model was initiated by Miller and Sanderson (2000; also see Miller, 2004) in the context of a study of intensive care units in hospitals. They argued that this type of model is not suitable for representing biological systems such as the human body. Their rationale was that control systems (e.g., bioregulatory mechanisms) are embedded in the human body and cannot be removed from a model of a patient, which suggests that they believe that control systems cannot be included in a work domain model.

Lind (2003) disagrees with this position. He attributes Miller and Sanderson's (2000) view that control systems cannot be represented in a work domain model to two main factors. One is that Rasmussen's (1986) treatment of the representation of control systems in this kind of model is superficial. The other is that Vicente (1999) implies that control systems cannot be incorporated into this type of model by stating that work domain analysis is concerned with the functional structure of "The System being controlled, independent of any particular Worker, Automation, Event, Task, Goal, or Interface" (p. 10).

While Rasmussen (1986) is unclear about the representation of control systems in a work domain model, he and his colleagues include such components in many of their models. For example, Rasmussen (1986) lists feedback loops, which are necessary for control systems to function, in models of a manufacturing system[*] and a computer system. Furthermore, Rasmussen et al. (1994) include people, who exert control over other components of a system, in models of a library,[†] a hospital,[‡] and a city.[§] Similarly, Vicente and his colleagues represent people in models of the game of baseball[¶] (Vicente and Wang, 1998); emergency ambulance dispatch management[**] (Hajdukiewicz et al., 1999, adapted from Rasmussen et al., 1987); engineering design[††] (Burns and Vicente, 1995, 2000); and military command and control[‡‡] (Hajdukiewicz et al., 1999). On this basis, it may be argued that neither Rasmussen (1986) nor Vicente (1999) preclude the representation of control systems in a work domain model.

Nevertheless, it should be acknowledged that certain statements by various analysts, including the above quotation from Vicente (1999), seem to suggest that control systems cannot be incorporated into a work domain model. Hajdukiewicz (1998), for example, mentions that he modified his abstraction–decomposition space of the human body to remove control from the

---

[*] See Figure 3.2 for the model of the manufacturing system.
[†] See Figure 3.11 for the model of the library.
[‡] See Figure 6.3 for the model of the hospital.
[§] See Figure 7.1 for the model of the city.
[¶] See Figure 3.10 for the model of the game of baseball.
[**] See Figure 5.6a for the model of the emergency ambulance dispatch management system.
[††] See Figure 6.1 for the model of the engineering design system.
[‡‡] See Figure 6.8 for the model of the military command-and-control system.

model.* In addition, Jamieson and Vicente (2001) state that, "While control systems are crucial to the successful operation of a modern petrochemical plant, they do not lend themselves to characterisation by structural means–ends descriptions" (p. 1060). Hence, control systems were not represented in their abstraction–decomposition space of a simulation of a fluid catalytic cracking unit in a petrochemical refinery[†] (Jamieson, 1998; Jamieson and Vicente, 2001).

On examining these articles in detail, however, it becomes apparent that analysts' concerns are specifically about representing the *behavior* of control systems in a work domain model. For instance, in the context of his study of the human body, Hajdukiewicz (1998) points out that one difference between conventional medical representations and a work domain model is that the former "incorporate control activity affecting system behaviour" and "mix system behaviour with structure" (p. 66). Likewise, in the context of their study of a simulation of a petrochemical system, Jamieson and Vicente (2001) state that "Other analysis techniques can be employed to model the behavior of control systems" (p. 1060).

Excluding the behavior of control systems from a work domain model is consistent with the theoretical orientation of work domain analysis. As mentioned before, the point of this technique is to create an event-independent representation of the constraints on actors' behavior so that it becomes possible to produce designs that assist actors in dealing with an assortment of situations, including those that are unforeseen. For this reason, a work domain model is concerned solely with the objects of action and structural means–ends relations. By portraying the behavior of control systems in such a model, analysts would be incorporating actions and action means–ends relations into the representation.

While describing the behavior of control systems in a work domain model is undesirable, I argue that it is appropriate to represent the structural properties of control systems. As stated previously, work domain analysis models the functional structure of actors' environments, including physical, social, or cultural properties, which place constraints on their behavior. If the structural properties of control systems, such as their functional capabilities or limitations, impose constraints on actors, then these properties should be incorporated into a work domain model. Otherwise, it would not be possible to develop designs using work domain analysis that support actors in reasoning about these constraints.

The position that it is possible to represent the structural properties of control systems in a work domain model, but not their behavior, is substantiated by those models incorporating control systems cited previously. These are models of a manufacturing system and a computer system (Rasmussen, 1986); a city, a hospital, and a library (Rasmussen et al., 1994); a military command-and-control system as well as an emergency ambulance dispatch management system (Hajdukiewicz et al., 1999); an engineering design system (Burns and Vicente, 1995, 2000); and the game of baseball (Vicente and Wang, 1998). Although these work domain models include the structural properties of control systems, they do not represent their behavior.

One further point is worth making in the context of this discussion. In considering the treatment of automation in work domain analysis, Burns et al. (2004) suggest creating dual models of a system. The first model would represent the system to be controlled. The second model would be the same as the first except that it would include automation or components of the system that provide control. Judging from Burns et al.'s comments about the treatment of sensors in work domain analysis, it seems that the main reason for their suggestion is that the first model would provide explicit support for defining requirements for

---

* See Figure 3.8a for the model of the human body.
† See Figures 4.1a and 4.1b, 4.8, and 4.9 for models of different aspects of the simulation of a fluid catalytic cracking unit in a petrochemical refinery.

automation design or redesign. That is, by focusing solely on the system to be controlled, the first model would offer analysts a framework for systematically identifying aspects of the system that are candidates for automation. The second model, which would be created following some level of specification of the automation design, would highlight interactions between the automation and other components of the system. These interactions may also be important to take into account during design, not just of the automation, but also of other aspects of the system, such as its displays.

A case study that illustrates an approach to modeling automation similar to that described above was conducted by Mazaeva and Bisantz (2007), who formulated dual models of a camera. The first model represents the camera without its automated components (Figure 7.2), while the second represents just the automated components of the camera (Figure 7.3). The models of the camera and its automation can be juxtaposed to highlight the interactions between them (Figure 7.4). As discussed by Mazaeva and Bisantz, these

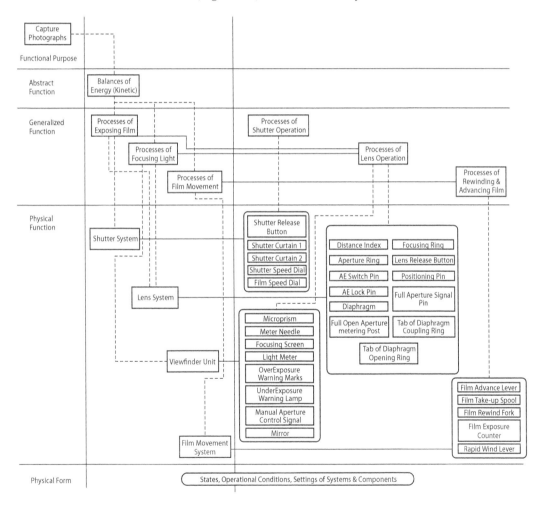

FIGURE 7.2
Abstraction–decomposition space of a camera without its automated components. (From Mazaeva, N., and Bisantz, A. M., 2007, On the representation of automation using a work domain analysis, *Theoretical Issues in Ergonomics Science*, *8*(6), 509–530. Taylor & Francis, reprinted by permission of the publisher [Taylor & Francis Ltd., http://www.tandf.co.uk/journals]; adapted with permission of the authors.)

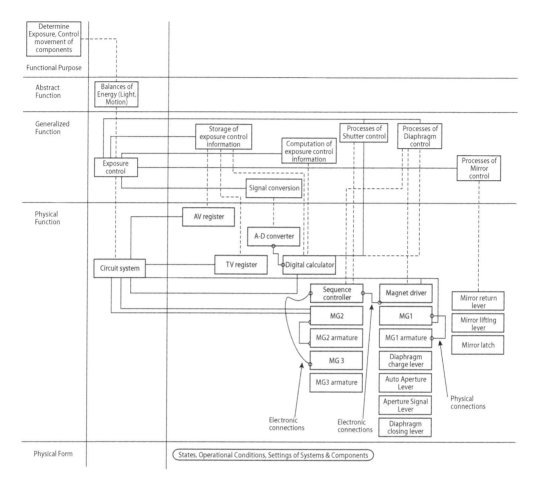

FIGURE 7.3

Abstraction–decomposition space of the automated components of a camera. (From Mazaeva, N., and Bisantz, A. M., 2007, On the representation of automation using a work domain analysis, *Theoretical Issues in Ergonomics Science*, 8(6), 509–530. Taylor & Francis, reprinted by permission of the publisher [Taylor & Francis Ltd., http://www.tandf.co.uk/journals]; adapted with permission of the authors.)

interactions show the functions, subsystems, and components of the camera that are controlled by the automation.

Dual models of a system assembled to highlight distinctions between the components to be controlled and those that provide control may be viewed as a form of multiple models. However, this style of representation was not covered in the preceding chapter because research in this area is still at an early stage.

Is There an Overlap between Work Domain Analysis and Control Task Analysis?

The final issue that this chapter addresses is whether there is an overlap between work domain analysis and control task analysis, which is concerned with the modeling of

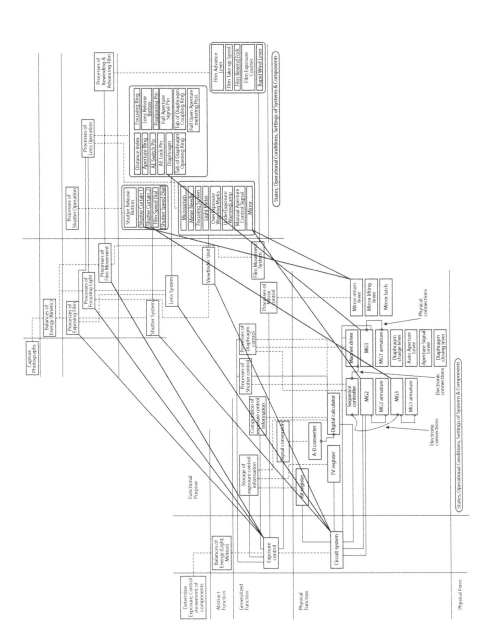

FIGURE 7.4

Interactions between the models of a camera and its automation. (From Mazaeva, N., and Bisantz, A. M., 2007, On the representation of automation using a work domain analysis, *Theoretical Issues in Ergonomics Science*, *8*(6), 509–530. Taylor & Francis, reprinted by permission of the publisher [Taylor & Francis Ltd., http://www.tandf.co.uk/journals]; adapted with permission of the authors.)

activity. As both of these dimensions of cognitive work analysis typically involve the modeling of functions, the question arises as to whether there are commonalities between the two dimensions and, if so, what the precise nature of this overlap is.

As discussed previously, work domain analysis focuses on the constraints that the functional structure of actors' environments imposes on their behavior. Along with other kinds of constraints, a work domain model may include the functions that are afforded by a system's physical resources. These functions, which are necessary for fulfilling a system's purposes, are referred to here as purpose-related functions, the label that Rasmussen most recently assigned to this category of constraint (Reising, 2000).

Control task analysis, on the other hand, focuses on modeling the activity that is necessary in a system to fulfill its purposes, given a suite of physical resources. This activity may be characterized initially as a set of recurring work situations or work functions (Rasmussen et al., 1994), using the contextual activity template[*] (Naikar et al., 2006). Following that, the activity may be characterized as a set of recurring control tasks for each work situation or work function (Rasmussen et al., 1994). The decision ladder template is commonly employed for modeling control tasks relating to the observation of information, situation analysis, goal evaluation, planning, and execution.[†]

An overlap between these two dimensions may be apparent, particularly when a work domain model includes purpose-related functions and a contextual activity template represents work functions. This may be demonstrated in the case of a home. The abstraction–decomposition space of a home in Figure 2.7 contains purpose-related functions such as *preparation of meals and beverages* and *rest*. A contextual activity template of the same home[‡] may include work functions such as 'prepare meals and beverages' and 'rest.' The two models, therefore, can appear to represent the same functions. This overlap between the two dimensions may seem more profound if the purpose-related functions in a work domain model are labeled with verbs. Given that verbs tend to signify activity, the work functions in a contextual activity template are likely to be represented in this form.

This overlap between work domain analysis and control task analysis does not mean that they model the same constraints. Instead, the overlap indicates a logical relationship between the two dimensions. A work domain model may include functions, or purpose-related functions, which are *afforded* by a system's physical resources. In contrast, a contextual activity template may contain functions, or work functions, which need to be *performed* in the system. The functions afforded by a system's physical resources are necessary for fulfilling its purposes. It follows, therefore, that the functions that must be performed in the system will correspond with those afforded by the physical resources. Nevertheless, the functions modeled by work domain analysis and control task analysis are theoretically distinct.

Summary

Earlier texts leave in doubt several matters concerning the relationship between the analysis of activity and work domain analysis. This chapter attended to four such matters. First,

---

[*] See Figure 1.5a for the contextual activity template.
[†] See Figure 1.5b for the decision ladder template.
[‡] See Figure A.1 in the Appendix for a hypothetical contextual activity template of a home.

it clarified that a work domain model should not represent actions and action means–ends relations but, instead, represent the objects of action and structural means–ends relations. Second, it recommended that analysts use nouns as often as possible for labeling the constraints in a work domain model. While analysts should not rule out the use of verbs entirely, they need to be careful that activity is not incorporated into the model accidentally. Moreover, analysts may find it useful to develop a glossary that describes the constraints in a work domain model in more depth, regardless of whether these are labeled with nouns or verbs. Third, this chapter established that it is possible to represent the structural properties of control systems in a work domain model, but not their behavior. Finally, it clarified that although both work domain analysis and control task analysis may involve the modeling of functions, this does not mean that they are concerned with the same constraints. Instead, this overlap points to a logical relationship between the two dimensions. Specifically, the purpose-related functions in a work domain model are functions *afforded* by a system's physical resources, while the work functions in a contextual activity template are functions that need to be *performed* in the system, which naturally correspond to those afforded by the physical resources.

This chapter draws to a close Section II of this book (Chapters 3 to 7), which examined the basic concepts of work domain analysis in some depth. The next part, Section III (Chapter 8), presents guidelines for performing work domain analysis, drawing on the material in previous parts.

Section III

Guidelines

8

Analytic Themes for Work Domain Analysis

OVERVIEW   This chapter presents guidelines for performing work domain analysis, not in the form of a series of steps, but in the form of a set of analytic themes or questions. These themes are as follows: (1) What is the purpose of the analysis? (2) What are the project restrictions? (3) What are the boundaries of the analysis or what is the focus system for the analysis? (4) Is it useful to develop multiple models? (5) Where on the causal–intentional continuum does the focus system fall? (6) What are the sources of information for the analysis? (7) What is the content of the abstraction–decomposition space? and (8) Is the abstraction–decomposition space a valid model of the focus system? A thorough explanation of these analytic themes is provided, including why each one is important, and suggestions or strategies for addressing each theme are offered.

Introduction

The guidelines for work domain analysis provided here are deliberately structured, not as a series of steps for analysts to execute, but as a set of analytic themes for analysts to contemplate. This format is significant because it reflects the fact that the process of performing this kind of analysis is flexible and iterative rather than rigid and linear. Precisely how an analysis is done depends a great deal on contextual factors, such as the intended application of the work domain model or even the project's schedule and budget, which may change over time. It is also rare for any part of the analysis to be dealt with or completed in a single attempt. Instead, each aspect is likely to be reconsidered and refined repeatedly, often as more knowledge of the system is gained. For instance, analysts may form a general idea of the purposes and boundaries of an analysis, visualize or sketch the likely content of the abstraction–decomposition space in light of these factors, and then revise the purposes and boundaries, having learnt more about the system during this modeling process. In addition, different aspects of the analysis are often seemingly considered in parallel rather than in a serial fashion. For these reasons, it is both undesirable and difficult—if not impossible—to formulate a prescriptive methodology for work domain analysis.

   The analytic themes were conceived on the basis of several distinct considerations. Of primary importance were the basic concepts of work domain analysis, which were introduced in Chapters 1 and 2 and addressed in depth in Chapters 3 to 7. The guidelines place these fundamental concepts within a methodological context.

   The analytic themes were also informed by numerous case studies of the application of work domain analysis, including my own. For example, I have used this technique to evaluate competing design concepts for an Airborne Early Warning and Control system (Chapter 9), to develop a team design for this system (Chapter 10), and to examine training needs and instructional system requirements for the F/A-18, a fighter aircraft (Chapter 11).

As shown in the next part of this book, my approach to modeling these large-scale military systems in industrial settings was useful and feasible.

Last, the analytic themes were shaped by evaluations of earlier versions of the guidelines. The first version, which was reported by Naikar, Hopcroft, and Moylan (2005), was assessed through its application to an advanced fighter aircraft (Hopcroft and Naikar, 2005); a manned space system (Baker, 2006; Baker, Naikar, and Neerincx, 2008); and a commercial training helicopter (Baker, 2006). The main outcome of these evaluations was a shift from portraying the guidelines as a series of steps to a set of analytic themes. Assessments of two later versions of the guidelines were made on the basis of their application to an uninhabited aerial system (Elix and Naikar, 2008c); a maritime surveillance system (Elix and Naikar, 2008b); an air power system (Lambeth and Naikar, 2008; Yeung and Naikar, 2010); and an air combat system (Treadwell and Naikar, 2009). These studies resulted primarily in improvements to the usability of the guidelines.

The rest of this chapter describes the eight analytic themes that constitute the guidelines for work domain analysis (Figure 8.1). These themes, which are intertwined and difficult to tease apart in practice, represent the major recurring concerns of analysts in developing a work domain model. The various questions may be considered in any order, and several problems may be under deliberation simultaneously. Thus, the themes may be visualized as bubbles that surface and fade continually in any sequence or combination.

The analytic themes are numbered in a rationalized order, which may be informative for inexperienced analysts. The first five themes involve decisions that influence the nature of the abstraction–decomposition space that is created. The sixth and seventh questions

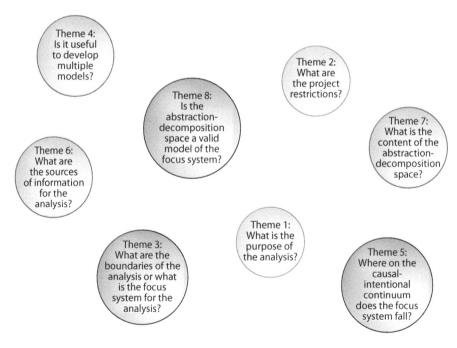

FIGURE 8.1
Analytic themes for work domain analysis.

concern the actual content of the model and how this may be developed. The last theme relates to the comprehensiveness and accuracy of the resulting representation.

The guidelines are applicable irrespective of whether analysts choose to model a work domain with an abstraction–decomposition space or an abstraction hierarchy. As highlighted in Chapter 4, the constraints in an abstraction hierarchy do fall at some level of decomposition, although this is not apparent in this style of representation.

When it is useful, the analytic themes are illustrated with examples from a home. These examples are not drawn from a single case study because it is difficult to devise a particular situation for demonstrating all aspects of the guidelines effectively. Instead, multiple hypothetical analyses of a home are assumed. The application of the analytic themes in three specific cases is described in the following chapters. As these cases deal with large-scale military systems in industrial settings, they provide more complex demonstrations of the guidelines.

Novices may find it useful to *initiate* the development of a work domain model by working through the analytic themes in numerical order. As indicated earlier, however, they should not expect to finalize their thinking in relation to each question on the first attempt but, rather, expect to reconsider or refine their ideas many times. I have also observed that beginners usually find it efficient to keep written notes of their thoughts or decisions regarding each theme, which can be easily reviewed and revised. Performing a work domain analysis requires a complex combination of activities that are difficult to juggle mentally without any aids or props, especially when one is new to the process. As analysts gain expertise in this technique, though, they may find themselves addressing many of the questions instinctively rather than deliberately.

In the following sections, each analytic theme is explained in depth, and the essential points are reiterated in boxed text. Sometimes, the text boxes impart additional advice as well. This means that after readers have become familiar with the details of each theme, they may rely predominantly on the text boxes for guidance.

Theme 1: What Is the Purpose of the Analysis?

A key question for analysts to consider in performing a work domain analysis is, What is the purpose of the analysis? Establishing the purpose is essential because there are many different ways to model the same system, and different representations of a system are useful for different applications. Thus, how analysts choose to model a system will depend greatly on the purpose.

Vicente (1999) emphasizes this point by highlighting that different types of maps of the same geographical area are appropriate for different purposes. So, whether a map of landscape elevation or average annual rainfall is necessary is contingent on one's purposes. Similarly, as Rasmussen (1986) points out, different kinds of maps of a city are effective depending on whether one's purpose is town planning or sightseeing.

Establishing the purpose of an analysis involves two important considerations. One is what objective will be achieved or what problem will be addressed with the analysis. The other is how a work domain model will be used to attain that objective. It is only by taking these factors into account that it is possible to create a model that is suitable for the intended application.

If the objective of an analysis is to design an ecological interface for a system, the process of using a work domain model to achieve the desired end is fairly well specified (e.g., Burns and Hajdukiewicz, 2004; Rasmussen and Vicente, 1989; Reising and Sanderson, 2002a; Vicente and Rasmussen, 1990, 1992). There are also reasonably detailed case studies that show how a work domain model can be used to evaluate alternative design concepts (Chapter 9), develop team designs (Chapter 10), examine training needs and instructional system requirements (Chapter 11), and formulate software specifications (Leveson, 2000). For an innovative application, however, the process of using a work domain model to address the problem will need to be planned.

The purpose of an analysis shapes numerous decisions relating to the type of model to be constructed. These include decisions about where to draw the boundaries of the analysis, whether it is useful to create multiple models, what levels of abstraction and decomposition to incorporate into the model, which cells of the abstraction–decomposition space to populate with constraints, what level of granularity or detail to represent the constraints at, and so on. Making such decisions without due regard for the intent of the analysis greatly increases the risk that the resulting representation will not be suitable for the proposed application.

To demonstrate how the purpose of an analysis can influence the type of model that is composed, the rest of this section presents three simplified case studies of a home. The first case is the development of a work domain model of a home for the purpose of exemplifying the ideas in this book. Specifically, the objective was to illustrate this material with a familiar system in which actors' behavior is shaped primarily by social laws, conventions, or values rather than physical or natural laws. The basic plan for using the resulting model (Figure 2.7) to fulfill this objective was that its levels of abstraction and decomposition, constraints, and structural means–ends and part–whole relations would be drawn on as examples. Given this purpose, the boundaries of the analysis were defined such that the home was classified as an intentional rather than a causal system.[*] In addition, partial abstraction hierarchies[†] were created for several of the rooms or subspaces and contents of the home at the second and third levels of decomposition in the model, so that a breadth of examples was made available. It was also decided that some of the constraints would be represented at a relatively coarse level of granularity because a more detailed model would have been unnecessarily complex for the purpose. For example, instead of describing the full set of functional capabilities or limitations of a *dishwasher*, which would have included its load limits, washing cycles, and temperature range, the object-related processes of this component were represented simply as *washing capacity*.

Alternatively, consider the case in which the objective of a work domain analysis of a home is to design an ecological interface for a tool that will assist inhabitants with meal preparation. To achieve this aim, it may be decided that the constraints in the resulting model will be used to derive inhabitants' information requirements, and the structural means–ends, part–whole, and topological relations in the model will be used as a basis for organizing this information on a display. In view of this purpose, analysts may choose to focus solely on the problem of meal preparation in the home. Accordingly, the boundaries of the analysis may include only those rooms or subspaces where meals are prepared, which may be just the kitchen and the yard (assuming the yard has a barbecue). Furthermore, the constraints may be represented at a level of granularity that allows vari-

---

[*] See Chapter 2 for an explanation of the distinction between intentional and causal systems.
[†] Partial abstraction hierarchies are discussed in Chapter 4.

> **BOX 8.1   WHAT IS THE PURPOSE OF THE ANALYSIS?**
>
> - What objective do I want to achieve with work domain analysis?
> - How will I use a work domain model to fulfill that objective?
>
> **HINTS**
>
> - The processes for using a work domain model to design ecological interfaces, evaluate design concepts, develop team designs, examine training needs and instructional system requirements, and formulate software specifications are reasonably well specified.
> - For an innovative application, analysts will need to define how they will use a work domain model to fulfill their objective.

ables to be extracted from each category, which can be used for measuring and displaying information about those constraints to inhabitants.

Consider the final case in which the reason for performing a work domain analysis of a home is to determine how to renovate the house in question so that it best supports the lifestyle of the inhabitants. To this end, it may be decided that the higher levels of abstraction in the resulting model will be used to define the inhabitants' purposive lifestyle requirements, and the lower levels of abstraction will be used to define their physical lifestyle requirements. Such a model could then be used to examine the impact of potential renovations to the house on the inhabitants' lifestyle.* If the aim of the model is limited to establishing the scope of internal renovations to the house, the boundaries of the analysis may exclude external areas such as the yard, the garage, and the driveway. In addition, if the renovations could involve structural alterations to the house, the boundaries may be delineated such that the home is classified as a system with a greater degree of causal constraints compared with the home modeled in Figure 2.7.

Theme 2: What Are the Project Restrictions?

Another question for analysts to think about in undertaking a work domain analysis is, What are the project restrictions? This is an important theme because, like the purpose of the analysis, these restrictions can shape how the analysis is done.

Generally, project restrictions are related to schedule, personnel, and finances. *Schedule* refers to the time that is available, for both doing the analysis and applying the resulting model to arrive at some end, for example, the implementation of an ecological interface or a team design. *Personnel* concerns the number and experience of the staff that are available, and *finances* relates to the amount of money that is available.

The project restrictions can shape how a work domain analysis is performed in a variety of ways. First, they may affect decisions about the purpose of the analysis itself. The tighter the restrictions, the narrower the purpose may be. Second, they may shape decisions

---

* A comparable application of work domain analysis for a military system is described in Chapter 9.

BOX 8.2   WHAT ARE THE PROJECT RESTRICTIONS?

- What is the nature of the project restrictions?
- How should I perform the work domain analysis given those restrictions?

HINTS

- Project restrictions are generally related to schedule, personnel, or finances.
- These restrictions may shape the purpose of the analysis, the scale of the model to be constructed, and the comprehensiveness of the process for performing the analysis.

about the scale of the model to be assembled, including how wide the boundaries of the analysis are drawn, whether multiple models are developed, how many parts for which full or partial abstraction hierarchies are created at each level of decomposition, and how many cells of the abstraction–decomposition space are populated with constraints. The more stringent the restrictions, the less extensive the scale of the model may be. Third, the restrictions may influence decisions about the comprehensiveness of the analysis process, for example, how many documents are reviewed, how many subject matter experts are interviewed, what length of walkthroughs or talkthroughs are undertaken, or how many scenarios are included in the validation exercise. The tighter the restrictions, the less comprehensive the process may be.

This theme, then, involves two main concerns. One is identifying the nature of the project restrictions, and the other is planning how the analysis will be performed in light of those restrictions. The latter requires judgments of what is feasible with respect to the purpose of the analysis, the scale of the model to be developed, and the comprehensiveness of the process for conducting the analysis.

Reconsider the case in which a work domain analysis of a home is required to establish what renovations are necessary to the house so that it suits the lifestyle of the inhabitants. In assessing the project restrictions, it may be ascertained that the owner of the home only has sufficient funds for renovating a few rooms; this is an example of a financial restriction. It may also be discovered that there is limited time for carrying out the analysis before the renovations are due to commence, which is an example of a schedule restriction. Given these restrictions, the purpose of the analysis may be limited to resolving what renovations to make to some rooms of the house, as opposed to the entire house. Consequently, the boundaries of the analysis may be placed around the most frequently used rooms or the rooms causing inhabitants the greatest dissatisfaction. Furthermore, because of schedule restrictions, the number and length of knowledge-elicitation sessions conducted with inhabitants may be moderated.

Theme 3: What Are the Boundaries of the Analysis?

An important question for analysts to address in performing a work domain analysis is, What are the boundaries of the analysis? This theme involves marking out the system, or those aspects of the system, that will be the focus of the study. Rasmussen et al. (1990) call this the *focus system*. The boundaries separate the focus system from its environment, so that the resulting model represents the focus system.

The decision about where to draw the boundaries of an analysis, or what should constitute the focus system, may be approached from several angles. Specifically, the boundaries may be determined by considering which *organizational entity* (e.g., home, restaurant, or hotel); *physical entity* (e.g., house, kitchen, or dishwasher); *problem* (e.g., meal preparation, cleaning, or house building); or *actors' perspective* (e.g., actors living in the home, visiting the home, cleaning the home, or building the home) should be the focus of the analysis. One or more of these considerations may be useful for demarcating the focus system.

Consider again the situation in which a work domain analysis of a home is required to design an ecological interface for a tool that will help inhabitants with meal preparation. The boundaries of this analysis may focus on the organizational entity of the home, the physical entities of the rooms or subspaces where meals are prepared (e.g., kitchen, yard), and the problem of meal preparation from the perspective of actors living in the home. Such an analysis would result in a model of the functional structure of the physical, social, or cultural environment of actors living in the home, principally as it relates to the problem of meal preparation.

Alternatively, reflect on the case in which the purpose of a work domain analysis of a home is to design an ecological interface for a tool to support inhabitants with the maintenance of a dishwasher. In this instance, the focus system may comprise the organizational entity of the home, the physical entity of the dishwasher, and the problem of maintenance from the perspective of actors living in the home. The outcome of such an analysis would be a model of the structural properties of the environment of actors living in the home, primarily as it relates to the problem of maintenance. Additional examples of how the focus system for an analysis may be specified in these terms, and how the definition of the focus system may affect the content of a model, are provided later in this chapter (in particular, see Theme 5).

As discussed in Chapter 6, the actors delineating the focus system for an analysis must be conceptualized in a way that is event independent. This may be achieved by describing the actors in a form that is generic or nonspecific. For instance, 'actors living in the home' is a generic category; irrespective of whom those particular people are, the constraints associated with living in the home are invariant. If specific people were chosen to characterize the focus system, the result would not be a work domain model but a model of those stakeholders' object worlds.

It is also worth noting that the problem that demarcates the focus system for an analysis is not the same as the purpose of the analysis. Drawing on a previous example, the purpose of an analysis may be to design an ecological interface for a tool to assist inhabitants with meal preparation. Meal preparation is the problem of interest for actors living in the home. These actors are not concerned with designing an ecological interface for a tool that supports meal preparation.

The decision of where to draw the boundaries of an analysis is mainly a pragmatic one. Generally, there are no actual or correct boundaries in the sense that there will always be elements outside the focus system that are coupled or linked to elements within the focus system (Rasmussen et al., 1990). Consequently, changes to external elements can propagate or spread to internal elements and vice versa. The boundaries, therefore, are essentially artificial divisions that are necessary for keeping the analysis within a manageable and useful scope (Burns et al., 2005).

Choosing the boundaries of an analysis involves two key considerations. First, it is necessary to take into account the purpose of the analysis and the project restrictions. A

BOX 8.3   WHAT ARE THE BOUNDARIES OF THE ANALYSIS?

- What is the focus system for the analysis? That is, which organizational entity, physical entity, problem, or actors' perspective should the analysis focus on?

HINTS

- Analysts need to take into account the purpose of the analysis, the project restrictions, and any natural demarcations in the system.
- Instances of proper names may signify natural demarcations in the system (Rasmussen, 1996).
- The actors who characterize the focus system for an analysis must be described in a way that is event independent.
- The problem that defines the focus system for an analysis is not the same as the purpose of the analysis.
- In weighing up what should be included in a work domain model, analysts may find it useful to consider what actors within the focus system need to reason about, or act on, given the problem with which they are concerned. Generally, these elements fall within the boundaries of the analysis.
- Burns and Hajdukiewicz (2004) suggest that if analysts are unsure about where to draw the boundaries of an analysis, it is better to err on the side of breadth.
- As discussed in Theme 4, if analysts choose to create multiple models of a system, there will be multiple focus systems for the analysis.

boundary that is too tight for the *purpose* will omit relevant features of the system, which could lead to designs that are limited or inadequate (Figure 8.2a). A boundary that is broader than it needs to be given the purpose will incorporate less significant features of the system, possibly with little gain in the quality of the resulting solutions (Figure 8.2b).

On the other hand, a boundary that is tighter than it needs to be in light of the *project restrictions* could result in excessive resources being spent on modeling particular features of the system at the expense of other relevant elements, perhaps with little benefit to the quality of the designs relating to those features (Figure 8.2c). A boundary that is too broad given the project restrictions could result in insufficient resources being allocated to modeling the most important features of the system, and thus lead to designs that are unsatisfactory or deficient (Figure 8.2d). Attempts to select a boundary that is suitable for both the purpose and the project restrictions may lead analysts to redefine the purpose or modify the project restrictions (e.g., by obtaining more time, personnel, or money).

Second, it is advisable to take into account any natural demarcations in the system so that the focus system for an analysis is as well bounded as possible. Rasmussen (1996) recommends that boundaries are drawn so that elements within the focus system are more tightly coupled to each other than they are to external elements. This means that perturbations to internal elements are more likely to propagate to other elements within the focus system relative to external elements. To offer a simple illustration, items of furniture within the bedroom of a house are more strongly coupled to each other than they are to furniture in other rooms. As a result, changing the position of a bed in the bedroom is more likely to affect the arrangement of other furniture in that room compared with furnishings in other

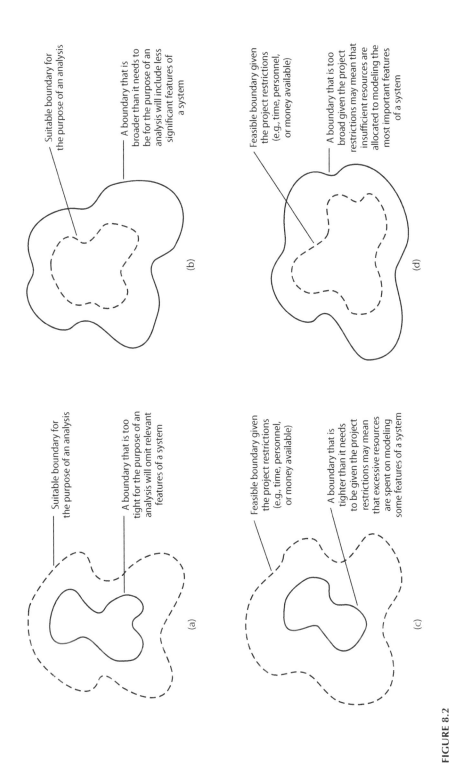

FIGURE 8.2

Defining the boundaries of an analysis involves taking into account the purpose of the analysis and the project restrictions: (a) a boundary that is too tight for the purpose; (b) a boundary that is broader than it needs to be for the purpose; (c) a boundary that is tighter than it needs to be given the project restrictions; (d) a boundary that is too broad in light of the project restrictions.

rooms. In view of these couplings, placing the boundaries of an analysis around the bedroom rather than half of that room will result in the focus system being better bounded. While placing the boundaries around the entire house may mean that the focus system is better bounded still, this may not be necessary for the purpose of the analysis or feasible given the project restrictions.

Rasmussen (1996) observes that groups of elements with tight coupling to each other usually have proper names. From this it follows that groups of elements with proper names may signal natural demarcations in a system. For example, the rooms of a house have proper names. In contrast, there is no proper name for one half of a bedroom. Like those of physical entities (e.g., house, kitchen, dishwasher), proper names for organizational entities (e.g., home, restaurant, hotel); problems (e.g., meal preparation, cleaning, house building); or actors (e.g., inhabitants, visitors, hired helpers, builders) may provide cues to natural boundaries in a system. Therefore, placing the boundaries of an analysis around groups of elements with proper names may help analysts ensure that the focus system is appropriately demarcated.

Theme 4: Is It Useful to Develop Multiple Models?

Also worthwhile contemplating in performing a work domain analysis is the question of whether it is useful to formulate multiple models of the same system. As discussed in Chapter 6, the term *multiple models* is used to refer to a series of abstraction hierarchies, or abstraction–decomposition spaces, created to capture different stakeholders' perspectives of a problem or to emphasize distinct facets of a problem. If the decision is taken to construct multiple models, there will be multiple focus systems for the analysis. In this sense, this theme relates to the boundaries of the analysis, but it is separated from the preceding discussion for the sake of clarity.

Multiple Stakeholders' Perspectives

One reason for constructing multiple models of the same system is to portray several stakeholders' perspectives of a problem. As noted in Chapter 6, *stakeholders* are individuals or groups with distinctive viewpoints of a specific problem or, in other words, distinctive *object worlds*. Given their overlapping, although different, concerns in the same problem, stakeholders normally have to interact or coordinate their activities to achieve their particular objectives. In addition, because of this coupling between their object worlds, changes or effects relating to one stakeholder can propagate to affect other stakeholders.

Creating multiple models of stakeholders' perspectives may be beneficial when designs intended for a particular stakeholder (or stakeholders) may have side effects on other stakeholders that should be taken into account during the development of the designs. In addition, multiple models of this type may be useful when designs are aimed at helping several stakeholders coordinate their activities better.

Consider, for example, the situation in which the purpose of a work domain analysis is to devise an ecological interface for a tool that supports meal preparation in a home. If the tool is intended just for inhabitants, only a single model may be produced to represent the problem of meal preparation from the perspective of actors living in the home. However, if the tool may also be used by visitors to help the inhabitants with meal preparation, it may be worthwhile to create a second model to depict this problem from the viewpoint of actors visiting the home. In addition, if the tool may have side effects on hired helpers, who clean the home, a third model may be beneficial to portray this problem from the perspective of actors cleaning the home. By developing multiple models to capture these different stakeholders' perspectives, analysts may be better placed to design an ecological interface that supports each stakeholder effectively and helps all stakeholders coordinate their activities without creating any undesirable side effects.

Figure 8.3 provides an outline of the three models that may be constructed to represent the problem of meal preparation from the perspectives of actors living in the home, visiting the home, and cleaning the home. The work domain of actors living in the home is labeled 'host,' whereas those of actors visiting and cleaning the home are called 'guest' and 'cleaner,' respectively. The ovals signify the object worlds of the three stakeholders in the problem of meal preparation, namely, inhabitants, visitors, and hired helpers. These stakeholders are portrayed as having shared, or overlapping, object worlds, mainly highlighting that they all may be concerned with cleaning in the home, although the full nature of the interactions between these stakeholders cannot be easily depicted in this figure. The text at the bottom of the figure shows one way in which the focus systems for this analysis may be defined, but many different ways are possible, depending on the exact nature of the system, the specific purpose of the analysis, and the project restrictions.

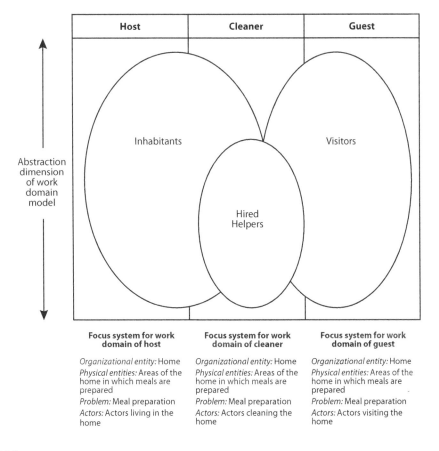

FIGURE 8.3
Multiple models of stakeholders' perspectives of the problem of meal preparation in a home.

Like Rasmussen et al.'s (1990) multiple models of stakeholders in a health care system,[*] the representation in Figure 8.3 differentiates between work domains (i.e., host, guest, and cleaner) and the object worlds of stakeholders (i.e., inhabitants, visitors, and hired helpers). As pointed out in Chapter 6, some of the advantages of this style of representation are that it can accommodate the possibility of (1) a stakeholder being concerned with more than one work domain; (2) several stakeholders being concerned with the same work domain, although with different but shared or overlapping aspects; and (3) stakeholders' object worlds being shared at a variety of levels of abstraction. Furthermore, this style of representation is consistent with the conceptual foundations of work domain analysis. Whereas work domains are event independent, stakeholders and their object worlds may change in response to the situational demands. For example, a family member may take over the role of hired helpers in a home, but the constraints associated with the work domain of cleaning remain the same. Preserving the distinction between work domains and stakeholders and their object worlds, therefore, is essential. As indicated previously, this may be achieved by ensuring that the boundaries of an analysis are independent of any specific actors (e.g., by focusing on a generic category such as 'actors cleaning the home'), even when one's intention is to capture the perspective of particular stakeholders (e.g., hired helpers).

Multiple Problem Facets

Multiple models of the same system may also be constructed to differentiate between the distinct facets of a problem that actors are faced with managing or to emphasize the basic differences in control that actors have over various elements of their work domain. As noted in Chapter 6, such models can then be used to formulate designs that appreciate these critical distinctions.

Normally, multiple models of problem facets are constructed for systems that are loosely bound (e.g., emergency management, military command and control) rather than tightly bound (e.g., process control). Actors in loosely bound systems do not have complete control over all of the entities or processes for which they have accountability, and they are responsible for handling disturbances that originate externally, as opposed to disturbances that stem from internal faults.

As pointed out in Chapter 6, by distinguishing between risks and resources for managing those risks, analysts have portrayed problems in loosely bound systems as having two distinct aspects. These are risk (or situation) assessment and risk (or resource) management. Thus, one aspect of such problems involves identifying, assessing, and prioritizing risks, whereas the other concerns identifying, assessing, and prioritizing resources for managing those risks.

Notably, analysts do not always develop multiple models of loosely bound systems. A single, integrated model is beneficial for emphasizing the interactions between the distinct facets of a problem or the various elements of a work domain. Designs that take into account these critical interactions may then be produced systematically.

Reflect on the case in which the purpose of a work domain analysis is to design a decision support tool to assist actors in an emergency management system to deal with emergencies in homes, such as fires, illnesses, injuries, or thefts. This system may be viewed as loosely bound because although its actors have control over the resources for managing

---

[*] See Figure 6.4 for the multiple models of stakeholders in the health care system.

emergencies (e.g., fire brigades, ambulances), they do not have control over the potential sources of emergencies (e.g., gas cookers, power tools) or the potential targets of damage (e.g., inhabitants, houses, neighborhoods). In addition, these actors have to respond to disturbances that arise externally (e.g., fires, injuries).

To design a decision support tool for this system, multiple models that highlight the distinct aspects of the problem of emergency management in homes may be useful. For instance, Figure 8.4a shows that two models could be developed by distinguishing between risks and resources. The model of the domain of potential risk would include the potential sources of emergencies and the potential targets of damage. The model of the domain of emergency management resources would include the resources for managing emergencies. The text at the bottom of the figure indicates how the focus systems for this analysis may be conceptualized.

Differentiating risks from resources for managing those risks allows the problem of emergency management in homes to be portrayed as having two distinct facets. The first, risk assessment, involves appraising emergencies in homes (domain of potential risk), and the second, risk management, involves evaluating resources for managing emergencies in homes (domain of emergency management resources). On the basis of this representation, a decision support tool could be devised that supports actors in assessing and managing risks more effectively. It is also worth noting that this representation distinguishes elements of the work domain over which the emergency management system has little or no control (domain of potential risk) from those over which it has greater control (domain of emergency management resources).

Alternatively, think about the situation in which the purpose of a work domain analysis is to evaluate alternative design concepts for the physical components of the emergency

Domain of Potential Risk	Domain of Emergency Management Resources		Emergency Management System
Includes the potential sources of emergencies (e.g., gas cookers, power tools) and the potential targets of damage (e.g., inhabitants, houses, neighborhoods)	Includes the resources for managing emergencies (e.g., fire brigades, ambulances)		Functional purposes, value and priority measures, and purpose-related functions of the emergency management system Physical objects and object-related processes of the emergency management system

Abstraction dimension of work domain model

Focus system for the domain of potential risk

Organizational entity: Emergency management system
Physical entities: Potential sources of emergencies and potential targets of damage
Problem: Risk assessment
Actors: Actors providing emergency management services

Focus system for the domain of emergency management resources

Organizational entity: Emergency management system
Physical entities: Resources for managing emergencies
Problem: Risk management
Actors: Actors providing emergency management services

Focus system for the emergency management system

Organizational entity: Emergency management system
Physical entities: Physical components of emergency management system
Problem: Emergency management (risk assessment and risk management)
Actors: Actors providing emergency management services

(a)   (b)

FIGURE 8.4
(a) Multiple models of the distinct facets of the problem of emergency management in homes; (b) a single, integrated model of the emergency management system.

management system, such as its communication devices, sensors, and workstations.[*] For this application, it may be useful to construct a single, integrated model of the system (for the problem of emergency management in homes), which shows how its physical objects and object-related processes are connected to its purpose-related functions, value and priority measures, and functional purposes. Such a model could be used to trace the impact of alternative designs for the physical components of the system, represented at the lower two levels of abstraction, on the constraints at the higher levels. This would lead to an understanding of how well the alternative designs support the management of emergencies in homes, including risk assessment and risk management.

Figure 8.4b provides an outline of the single model. If this model was completed, lower levels of abstraction would depict the physical objects and object-related processes of the emergency management system, and higher levels of abstraction would portray its functional purposes, value and priority measures, and purpose-related functions. A description of the focus system for this analysis is shown at the bottom of the figure.

Unlike the multiple models of the emergency management system (Figure 8.4a), the single model does not separate the representation of the potential sources of emergencies and the potential targets of damage from that of the resources for managing emergencies. Instead, these elements of the work domain would be integrated implicitly into the representation of the physical objects, object-related processes, purpose-related functions, value and priority measures, and functional purposes of the system. For instance, the resources for managing emergencies might be reflected in the object-related processes of the communication devices as these devices would need to be capable of supporting the exchange of information with those resources. Likewise, the potential sources of emergencies and the potential targets of damage might be manifested in the purpose-related functions of the system as it would need to be capable of assessing and managing the risks posed by those sources and targets.

The single model, then, still conveys that the problem of emergency management in homes concerns risk assessment and risk management. However, compared with the multiple models, the representation of the potential sources of emergencies, the potential targets of damage, and the resources for managing emergencies would be less explicit and less detailed in the single model. In contrast, the interactions between these elements would be more apparent in the single model than in the multiple models.

Finally, it is worth pointing out that both options may be beneficial for any given application. For example, the single model could be useful for designing a decision support tool for the emergency management system because it would reveal interactions between elements of the work domain that may be important for risk assessment or risk management. Multiple models, on the other hand, could be useful for evaluating alternative design concepts for this system because they would enable a more detailed evaluation of how well the designs support risk assessment and risk management. For some applications, analysts may decide to implement both options. However, if constrained by project restrictions, analysts will have to select the option that best satisfies their requirements.

---

[*] A case study of a similar application of work domain analysis for a military system is described in Chapter 9.

BOX 8.4   IS IT USEFUL TO DEVELOP MULTIPLE MODELS?

- Is it useful to develop multiple models to capture different stakeholders' perspectives of a problem?
- Is it useful to develop multiple models to emphasize the distinct facets of a problem or the fundamental differences in control that a system has over elements of its work domain?

HINTS

- Analysts may find it useful to create multiple models of stakeholders' perspectives when designs intended for a particular stakeholder (or stakeholders) may have side effects on other stakeholders or if designs are intended to help several stakeholders coordinate their activities more effectively.
- The boundaries of an analysis must not incorporate specific actors, even when one's goal is to portray stakeholders' perspectives.
- Multiple models of loosely bound systems may be useful for emphasizing the distinct aspects of a problem or the fundamental differences in control that a system has over elements of its work domain, so that designs that recognize these important distinctions may be developed systematically. On the other hand, a single, integrated model may be beneficial for emphasizing the interactions between the various facets of a problem or elements of a work domain, so that designs that recognize these significant interactions may be devised.
- If analysts decide to build multiple models of the same system, the analysis will comprise multiple focus systems.

Theme 5: Where on the Causal–Intentional Continuum Does the Focus System Fall?

Another question that is significant in performing a work domain analysis is, Where on the causal–intentional continuum does the focus system fall? As highlighted in Chapter 2, this is an important theme for analysts to take into account because it leads to an appreciation of whether intentional or causal constraints in the focus system are the *primary* influence on actors' behavior. Consequently, the nature of the constraints that should be represented in a model of that system may be established. Without this step, the model is more likely to be incomplete or inaccurate.

This theme, therefore, may be viewed as involving two essential considerations. First, it is necessary to examine the nature of the intentional and causal constraints in the focus system, so that the *approximate* location of the system on the causal–intentional continuum may be identified. The system's position relates to whether actors' behavior is governed mainly by intentional or causal constraints. Following that, an understanding of the nature of the constraints that should be portrayed in the model of the focus system, given its placement on the continuum, may be gained. To assist analysts with these activities, Figure 8.5 presents an amalgamation of Figures 2.13 and 2.14 from Chapter 2.

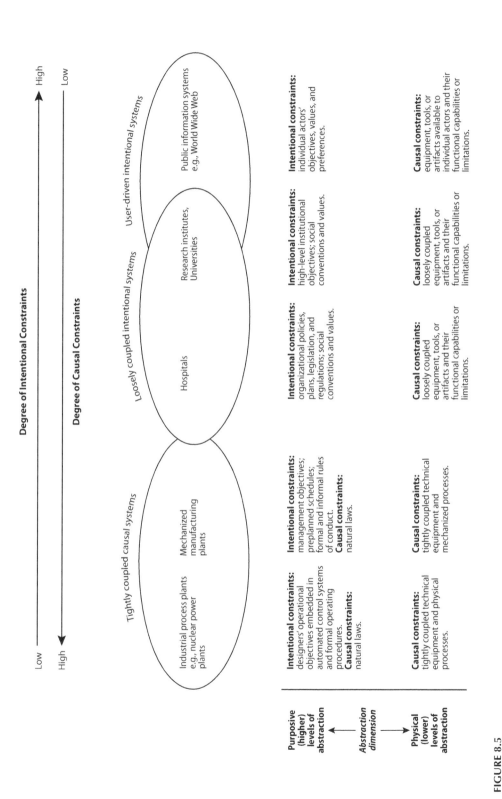

FIGURE 8.5

A representation for identifying where on the causal–intentional continuum the focus system falls and the nature of the constraints that should be represented in the model given the focus system's position.

To establish the nature of the intentional constraints in the focus system, the examples depicted in Figure 8.5 may be used as a guide. Specifically, moving along the causal–intentional continuum from left to right, analysts may consider whether intentional constraints in the focus system stem from (1) designers' operational objectives embedded in automated control systems and formal operating procedures; (2) management objectives, preplanned schedules, and formal and informal rules of conduct; (3) organizational policies, plans, legislation, and regulations; (4) high-level institutional objectives; or (5) individual actors' objectives, values, and preferences. In broad terms, workers have greater flexibility in setting goals and priorities and in organizing their daily activities as one proceeds along the continuum in this direction.

Similarly, Figure 8.5 may be used as a basis for establishing the nature of the causal constraints in the focus system. That is, analysts may reflect on whether causal constraints in the focus system are evident in (1) tightly coupled technical equipment and physical processes governed by natural laws; (2) tightly coupled technical equipment and mechanized processes governed by natural laws; (3) loosely coupled equipment, tools, or artifacts and their functional capabilities or limitations; or (4) equipment, tools, or artifacts available to individual actors and their functional capabilities or limitations. Generally, as one advances along the continuum from left to right, workers have greater control over how physical resources are employed, and production depends less on the functions of the technical equipment and more on the activities of workers.

By developing an appreciation of the nature of the intentional and causal constraints in the focus system, analysts will gain insight into where on the causal–intentional continuum that system broadly falls. It is worth noting that the focus system may be characterized by different types or combinations of intentional or causal constraints compared with those characterizing the systems in Figure 8.5. Nevertheless, this figure can still promote a general sense of the nature of the constraints in the focus system, which is all that is necessary for illuminating its approximate location on the causal–intentional continuum.

As stated previously, the focus system's position on the causal–intentional continuum signifies whether intentional or causal constraints are the principal determinant of actors' behavior. In Chapter 2, it was outlined that if the focus system falls toward the left of the continuum, so that it is a *tightly coupled causal system*, a high degree of causal constraints relative to intentional constraints shapes actors' behavior. Hence, actors' behavior is controlled predominantly by causal constraints. If the focus system falls toward the extreme right of the continuum, so that it is a *user-driven intentional system*, intentional constraints have more influence than causal constraints on actors' behavior. If the focus system falls toward the right of the middle of the continuum, so that it is a *loosely coupled intentional system*, causal constraints have greater bearing on actors' behavior than do such constraints in systems located at the far right. Nevertheless, in loosely coupled intentional systems, intentional constraints still carry more weight than causal constraints, so that actors' behavior is determined largely by intentional constraints (Rasmussen et al., 1994).

On the basis of an understanding of the focus system's location on the causal–intentional continuum, it becomes possible to establish the nature of the constraints that should be represented in a model of that system. Figure 8.5 summarizes the kinds of constraints that may be portrayed at purposive (higher) and physical (lower) levels of abstraction, given the system's position on the continuum. In the case of tightly coupled causal systems, the model may be dominated by causal constraints at both higher and lower levels. In addition, intentional constraints may be depicted at the higher levels. Some examples of models of tightly coupled causal systems are those of a microworld simulation of a thermalhydraulic system[*] (Bisantz and Vicente, 1994); the human body[†] (Hajdukiewicz, 1998; Sharp, 1996; Watson and Sanderson, 2007); a microworld simulation of a pasteurization plant[‡] (Reising and Sanderson, 2002b); and an automatic camera[§] (Mazaeva and Bisantz, 2007).

As user-driven intentional systems are characterized by autonomous, casual actors, it may be challenging to create a model that is universally relevant for actors in such systems, except in terms that are fairly broad (Rasmussen et al., 1994). Intentional constraints may be described mainly at higher levels of abstraction, while causal constraints may be portrayed solely at lower levels. An example of a model that falls in this category is that of a gymnastics computer game[¶] (Hansen et al., 1991).

In the same way, models of loosely coupled intentional systems may be characterized by intentional constraints predominantly at higher levels of abstraction and causal constraints exclusively at lower levels. Examples of such models include those of a home (Figure 2.7); a health care system[**] (Rasmussen et al., 1994); and a military command-and-control system for suppression of enemy air defense missions[††] (Rasmussen, 1998).

Finally, it is important to bear in mind that a system's position on the causal–intentional continuum is contingent on exactly how the focus system for an analysis is conceptualized. That is why the same system may be modeled as a causal system or an intentional system, depending on the definition of the focus system, as indicated in Chapter 2. For example, Rasmussen (1986) represented a manufacturing system much like a causal system in the sense that causal constraints were described at both higher and lower levels of abstraction in the model.[‡‡] In contrast, Rasmussen et al. (1990) portrayed a manufacturing system more like an intentional system in the sense that causal constraints were shown basically at lower levels.[§§] Similarly, Torenvliet et al. (2008) and Burns et al. (2005) have modeled naval command-and-control systems much like causal systems,[¶¶] whereas Linegang and Lintern (2003) and Bisantz et al. (2003) have depicted such systems more like intentional systems.[***] The following two cases of a work domain analysis of a home can be used to illustrate this point more clearly.

Consider the first case, in which the purpose of the analysis is to design a decision support tool to assist inhabitants with meal planning. The focus system for this analysis may comprise the organizational entity of the home, the physical entities of the rooms or subspaces where meals are prepared and served (e.g., kitchen, yard, and dining room), and the problem of meal planning from the perspective of actors living in the home. Given this focus system, it may be established that the intentional constraints on actors' behavior stem from the high-level objectives of the home (e.g., requirements for financial savings) as well as its social conventions and values (e.g., religious beliefs), which influence meal planning. The causal constraints have their basis in loosely coupled equipment, tools, or

---

[*] See Figure 2.15 for the model of the thermalhydraulic system.

[†] For representations of the human body, see Figure 3.7a for Sharp's model, Figure 3.8a for Hajdukiewicz's model, and Figure 5.4 for Watson and Sanderson's model.

[‡] See Figure 5.3 for the model of the pasteurization plant.

[§] See Figures 7.2, 7.3, and 7.4 for models of the automatic camera.

[¶] See Figure 2.16 for the model of the gymnastics computer game.

[**] See Figure 6.3 for the model of the health care system.

[††] See Figure 4.10 for the model of the military command-and-control system.

[‡‡] See Figure 3.2 for the representation of a manufacturing system as a causal system.

[§§] See Figure 3.6 for the representation of a manufacturing system as an intentional system.

[¶¶] For representations of naval command-and-control systems as causal systems, see Figure 6.5 for Torenvliet et al.'s model and Figure 6.9 for Burns et al.'s model.

[***] For representations of naval command-and-control systems as intentional systems, see Figure 6.6 for Linegang and Lintern's model and Figure 6.12 for Bisantz et al.'s model.

artifacts in the areas of the home used for preparing and serving meals, and the functional capabilities or limitations of those implements, which also influence meal planning. Accordingly, the focus system is located toward the right of the middle of the causal–intentional continuum, so that it is a loosely coupled intentional system. Intentional constraints have more bearing than causal constraints on actors' behavior. The model of the focus system may be characterized by intentional constraints mainly at higher levels of abstraction and causal constraints entirely at lower levels.

In the second case, the purpose of the analysis is to design an ecological interface for a tool to assist builders with the physical construction of a house. The boundaries of this analysis may be drawn around the organizational entity of the home, the physical entity of the house (including its foundations, roofing, electrical wiring, plumbing, etc.), and the problem of its physical construction from the perspective of actors building the home. Given this focus system, the intentional constraints on actors' behavior have their basis in the inhabitants' objectives, which are embedded in architectural specifications. The causal constraints stem from the tightly coupled materials of the house and their physical processes, which are governed by natural laws such as the law of gravity and the laws of conservation of mass and energy. The focus system, then, falls toward the left of the causal–intentional continuum, so that it is a tightly coupled causal system. A high degree of causal constraints compared with intentional constraints shapes actors' behavior. The model of the focus system may be characterized by causal constraints at both higher and lower levels of abstraction. Higher levels may portray constraints relating to natural laws, whereas lower levels may describe constraints relating to tightly coupled materials of the

---

BOX 8.5   WHERE ON THE CAUSAL–INTENTIONAL CONTINUUM DOES THE FOCUS SYSTEM FALL?

- What is the focus system's position on the causal–intentional continuum? That is, is actors' behavior shaped primarily by intentional or causal constraints?
- What is the nature of the constraints that should be represented in a model of that system?

HINTS

- To answer these questions, analysts may use Figure 8.5 as a guide to define (1) the nature of the intentional constraints in the focus system, (2) the nature of the causal constraints in the focus system, (3) the location of the focus system on the causal–intentional continuum (or whether intentional or causal constraints are the principal influence on actors' behavior), and (4) the nature of the constraints that should be portrayed in a model of the focus system.
- Analysts only need to develop a general idea of the nature of the intentional and causal constraints in the focus system, the location of the focus system on the causal–intentional continuum, and the nature of the constraints that should be depicted in a model of the focus system.
- A system's location on the causal–intentional continuum (which reflects whether actors' behavior is governed mainly by intentional or causal constraints, and therefore, the nature of the constraints that should be represented in a model) is highly dependent on how the focus system for an analysis is defined.

house and their physical processes. In addition, intentional constraints associated with inhabitants' objectives, as specified in architectural specifications, may be depicted at the higher levels.

Theme 6: What Are the Sources of Information for the Analysis?

A standard question for analysts to consider in performing a work domain analysis is, What are the sources of information for the analysis? The key information sources for this form of analysis are documents, field observations, and subject matter experts. These sources are useful for making a number of decisions relating to the preceding themes, such as what the boundaries of the analysis are, whether it is useful to develop multiple models, and where on the causal–intentional continuum the focus system falls. In addition, these sources are important for defining the content of the abstraction–decomposition space, which is the subject of the next theme. This section focuses on explaining how these sources may be utilized to address the next theme, particularly to define the specific constraints that should be represented in a work domain model. That is why this theme is presented at this point in this chapter; it was not presented prior to the earlier themes as this section does not address directly how the information sources can be used for addressing those themes.

Many readers will be aware that documents, field observations, and subject matter experts are customary sources of information for normative and descriptive approaches to work analysis.* In other words, these sources are commonly used for investigating how work should be done or is done by actors. It may seem puzzling, then, that the same sources can be used for formative approaches, that is, to establish how work can be done in a system. This issue may be settled by pointing out that the specific types of documents, field observations, and subject matter experts that are suitable for work domain analysis may be different from those required for normative or descriptive techniques. Moreover, even if some of the sources are relevant for all of the techniques, the information from these sources that is useful for work domain analysis is likely to be different from that which is necessary for the other techniques.

Specifically, unlike normative or descriptive techniques, work domain analysis takes advantage of documents, field observations, and subject matter experts to establish not *what* actors do or should do, but *why* some things must be done and the alternatives for *how* they can be done (Rasmussen et al., 1994). As explained previously, the aim of this technique is to develop a model of the functional structure of the physical, social, or cultural environment of actors. The structural properties of the environment place constraints on actors by specifying the fundamental reasons (why) and resources (how) for their behavior.

The three information sources rarely make explicit the constraints on actors' behavior. Instead, they are more likely to reveal how work is done or should be done in a system. For instance, they may describe the tasks or activities that workers normally carry out or the operating procedures that workers should follow. Nevertheless, by systematically considering the reasons and resources for actors' behavior, analysts can exploit documents, field observations, and subject matter experts for work domain analysis. That is, analysts can

---

* Chapter 1 discusses normative, descriptive, and formative approaches to work analysis.

utilize these sources to 'reverse engineer' or uncover the constraints on actors' behavior (Rasmussen et al., 1994).

An alternative way of expressing the preceding points, which may be instructive for readers, is that documents, field observations, and subject matter experts tend to provide information about the functioning of a system in specific instances instead of the constraints on actors' behavior, which are event independent. For example, these sources may emphasize the goals of a system in particular situations rather than its purposes, which remain relatively stable over time. Similarly, they may focus on a system's performance under certain conditions rather than the criteria it must satisfy to achieve its purposes irrespective of the circumstances. They may also emphasize how various tools are used to deal with different events instead of the functional capabilities or limitations of those implements, which do not vary as a function of the situation. Nevertheless, by assembling information about specific instances of system functioning, analysts can take advantage of documents, field observations, and subject matter experts to glean, or piece together, the constraints on actors' behavior.

The following sections explain in more detail how the three sources may be employed to gather information for work domain analysis. Examples are provided of the types of documents, field observations, and subject matter experts that are useful for this form of analysis but are not commonly used for normative or descriptive techniques. Where work domain analysis and the other techniques make use of the same types of sources in different ways, this is noted as well. These observations are important because they help to clarify how to make the best use of the three sources for work domain analysis.

The following sections also discuss the strengths and limitations of these sources for work domain analysis. Consideration is given to whether they are more useful during the earlier or later stages of the exercise. In addition, the suitability of these sources for providing information about a system's functioning during novel or unfamiliar situations and for modeling future systems is addressed. The former is important because the constraints on actors' behavior are likely to be more evident in challenging situations compared with those that are routine or familiar. As indicated in Chapter 2, in standard or recognizable circumstances, actors may not consider the constraints on their behavior explicitly (Rasmussen et al., 1994). The latter is important because basing the design of future systems on a comprehensive understanding of their work demands is critical for building safe, productive, and healthy systems. The strengths and limitations of these sources for work analysis in general are covered comprehensively elsewhere (e.g., Cooke, 1999; Kirwan and Ainsworth, 1992).

Documents

Documents are significant sources of information for work domain analysis. In fact, any documentation about the focus system may be worth reviewing, including organizational plans and reports, policy formulations and strategic directives, operating procedures and training manuals, textbooks and technical manuals, site maps and architectural drawings, engineering requirements and specifications, accident and incident reports, and articles in newspapers and magazines. Figure 8.6 shows that some kinds of documents may be more suitable than others for uncovering information about the focus system in relation to purposive versus physical levels of abstraction.

Different types of documents may be relevant for work domain analysis compared with normative or descriptive techniques. Staff or financial reports, for example, may be essential for developing a work domain model. Among other things, these documents may disclose

	Documents	**Field Observations**	**Subject Matter Experts**
Purposive (higher) levels of abstraction	Organizational plans and reports, Policy formulations, Strategic directives	Meetings for planning and reporting	Directors, Executives, Strategists, Legislators
Abstraction dimension	Operating procedures, Training manuals, Textbooks, Technical manuals	Front-line operations, Training sessions	Front-line workers, Training instructors
Physical (lower) levels of abstraction	Site maps, Architectural drawings, Engineering requirements and specifications	Maintenance operations	Maintenance personnel, Engineers

FIGURE 8.6

Types of documents, field observations, and subject matter experts that may be relevant for obtaining information about the focus system with respect to purposive versus physical levels of abstraction.

the personnel or monetary criteria for evaluating organizational performance, which may reflect the constraints on actors' behavior. Alternatively, some types of documents may be appropriate for all three techniques, but how these documents are employed for work domain analysis may be different from how they are utilized for the other techniques. As a case in point, the aim of examining a system's operating procedures for work domain analysis may be to uncover the fundamental laws, principles, or values that must be respected by actors rather than to translate the procedures into descriptions of task sequences.

Documents are extremely convenient information sources during the early stages of model construction, when analysts typically have limited understanding of the focus system. Using documents, analysts can develop considerable knowledge of the system at a pace that is suited to their level of understanding. Some kinds of documents may be informative with respect to how a system functions under novel or unfamiliar conditions. Accident or incident reports, for instance, may be revealing in this regard. Documents are also likely to be important for modeling future systems. Some examples of documents that may be useful for this purpose include concept or design papers, reports of cost–benefit assessments or performance predictions, and engineering requirements or specifications.

Field Observations

Field observations can also be instructive for work domain analysis. For example, it may be worth conducting observations of meetings for planning and reporting, frontline operations, training sessions, and maintenance operations. Different types of field observations may be useful for discovering information about the focus system with respect to purposive versus physical levels of abstraction (Figure 8.6).

The types of field observations that are appropriate for work domain analysis may be different from those used for normative or descriptive techniques. As a case in point, despite the fact that the focus system may be concerned with frontline operations, observations of executive planning meetings may be informative for modeling the work domain as such studies may shed light on the organizational objectives or values that impose constraints on actors' behavior. Although some kinds of field observations may be relevant for all of the techniques, there may be variations in how these field observations are utilized for work domain analysis versus the other techniques. For instance, the aim of observing training sessions for work domain analysis may be to identify the criteria that actors seek to fulfill in performing their jobs rather than to describe the details of their activities.

Field observations may be either exploratory or focused (Hajdukiewicz, Doyle, Milgram, Vicente, and Burns, 1998). Exploratory observations are conducted with the aim of developing a general knowledge of the focus system. Such investigations may be helpful during the early stages of model building, when analysts are still becoming familiar with the system. Focused observations, which are concerned with developing detailed knowledge of specific elements of the system, may be beneficial during the later stages, when analysts are formulating particular aspects of their models.

Field observations may be revealing about how systems operate in novel or unfamiliar situations if such events happen to occur naturally during periods of investigation or are introduced deliberately into training sessions. These kinds of incidents are fairly rare, however, which means that analysts will be limited mainly to observing how a system performs under standard conditions. Field observations are not possible for modeling future systems, at least during the initial stages of their development. Later during their development, mockups, prototypes, or simulations of future systems may provide analysts with opportunities for observation. It should also be noted that, unless a future system is highly similar to an existing one, field observations of existing systems may be worthwhile for learning about only the class of system to which a future system belongs.

Subject Matter Experts

Subject matter experts are also valuable information sources for work domain analysis. Typically, for modeling any focus system, a wide range of subject matter experts may be consulted, including directors, executives, strategists, legislators, administrators, managers, frontline workers, training instructors, maintenance personnel, scientists, engineers, and system developers. As depicted in Figure 8.6, people with different kinds of expertise may be useful for providing information about the focus system in relation to purposive versus physical levels of abstraction.

Compared with the subject matter experts consulted when using normative or descriptive techniques, those consulted for work domain analysis may be atypical. For example, even when the analysis is focused on frontline operations, engineers may be worth interviewing for work domain analysis in order to establish the technical capabilities or limitations of tools in the workplace, which place constraints on actors' behavior. As was true for the other information sources, some types of subject matter experts may be worth conferring with for all of the techniques, but the information from these subject matter experts that is necessary for work domain analysis may not be the same as that for the other techniques. For instance, in developing a work domain model, the aim of interviewing workers about their experiences in challenging situations may be to uncover the structural properties of the environment that shaped their decisions, not to analyze their cognitive strategies or processes.

With normative or descriptive techniques, the people who carry out the work that is under investigation are usually the main, or only, subject matter experts consulted for the study. For instance, for the analysis of frontline operations, frontline workers may be the sole informants for the study. While this approach may be quite appropriate for normative or descriptive techniques, it may be problematic for work domain analysis. One reason for this is that in stable or well-established systems, or under routine or familiar conditions, workers may not consider the constraints on their behavior explicitly (Rasmussen et al., 1994). Instead, they may simply invoke patterns of behavior that have been effective on prior occasions. These practices may even be passed on to successive generations of workers, so that the constraints that control behavior are no longer known to workers. Another problem is that many systems are too large and complex for workers to understand all aspects of their operation (Burns and Hajdukiewicz, 2004).

For work domain analysis, therefore, it is usually important to consult a variety of subject matter experts. With respect to the analysis of frontline operations, for example, all or many of the subject matter experts listed previously may be important. Collectively, these subject matter experts may have the range of knowledge or experience required for analysts to piece together the constraints on actors' behavior. For instance, while workers may provide insight into the regulations, standards, or operating procedures they follow, directors, executives, or legislators may shed light on the organizational objectives or values that shaped those instructions. Similarly, maintenance personnel, engineers, and system developers may reveal the full extent of the functionality of tools in the system, whereas workers may be limited to describing only the functionality they utilize regularly.

Many standard techniques for knowledge elicitation (see, for example, Cooke, 1999, and Kirwan and Ainsworth, 1992) are suitable for gathering information from subject matter experts for work domain analysis. In the following sections, I discuss the use of walkthroughs, talkthroughs, interviews, and tabletop analyses for this purpose. The strengths and limitations of these knowledge-elicitation techniques for performing this type of analysis may be readily extended to other techniques that are not reviewed here.

Walkthroughs and Talkthroughs

Walkthroughs and talkthroughs (Kirwan and Ainsworth, 1992) are two similar, but distinct, techniques. Walkthroughs involve experts performing demonstrations in realistic settings, whereas talkthroughs involve experts providing explanations away from such settings. Typically, both techniques have been applied to elicit information from workers about their tasks. In the case of walkthroughs, workers demonstrate their tasks at their actual workstations or in simulators, perhaps by pointing to displays, tools, or controls, but without actually performing their jobs. With talkthroughs, workers explain their tasks away from the workplace, so that their tasks are described rather than demonstrated.

For work domain analysis, the two techniques may be structured to extract information from experts about the constraints on actors' behavior. Workers, for instance, may still be asked about their tasks, but they may be requested to emphasize the reasons for performing those tasks and the resources necessary for supporting those tasks. Both techniques may be employed with a range of experts. For example, engineers may be asked to demonstrate or explain the technical capabilities and limitations of the tools available to workers.

The two techniques are most efficiently utilized once analysts have gained sufficient knowledge of the focus system to understand the information provided by experts without seeking extensive clarification of this material during the exercise. It is also important to point out that both techniques can be used for uncovering information about system functioning in novel or unfamiliar circumstances. Workers may be asked to provide an account of the reasons and resources for their actions in examples of these situations. Similarly, engineers may be required to identify the equipment functionality that is significant for handling an event of this type. However, in the case of both techniques, experts must be able to establish promptly, or without difficulty, how the system would function under those conditions.

One advantage of talkthroughs, compared with walkthroughs, is that they can be used for investigating future systems prior to their development. As indicated previously, this technique does not depend on experts having access to realistic settings. Hence, continuing with the themes of the preceding examples, experts may be asked to explain the reasons and resources for the tasks that are necessary in a future system or the technical capabilities and limitations of the tools that will be available in that system. Walkthroughs may be employed as soon as mockups, prototypes, or simulations of the future system become available. For both techniques, though, experts must have adequate knowledge of the future system to readily offer information relating to the constraints on actors' behavior.

Interviews

Interview formats may be described as unstructured, structured, or semistructured. Unstructured interviews are "free-flowing" discussions without the content or the sequence of the questions put to experts being defined in advance (Cooke, 1999, p. 487). In the case of structured interviews, at the other end of the spectrum, both the content and the sequence of the questions for experts are predetermined. Semistructured interviews fall in between these two extremes. In this format, some of the questions for experts are specified in advance, but the wording and the sequence of those questions are not fixed, and other questions may be introduced as necessary. Although structured interviews may be used for work domain analysis, unstructured and semistructured interviews are more common.

Rasmussen et al. (1994) describe an unstructured format for interviewing experts that involves asking them to identify the key functions in which they are engaged. After that, the reasons and resources for those functions or "means–ends relationships" can be examined (p. 56). For example, experts can be asked to discuss the objectives of a function, or the reasons it is performed, as well as the resources for that function, or how it can be achieved. Minor variations to this format include asking experts to concentrate on the functions they participate in during a single day or the functions that form part of a particular process.

A semistructured format based on the critical decision method is another option for conducting interviews with experts. The critical decision method relies on experts' accounts of nonroutine or challenging incidents in which they played a key decision-making role (Hoffman, Crandall, and Shadbolt, 1998; Klein, Calderwood, and MacGregor, 1989). Typically, in the course of an interview, a series of information-gathering sweeps are used to drill down to an expert's cognitive strategies or processes during a critical incident. For work domain analysis, this procedure may be adapted to target information about the constraints on actors' behavior. At least four sweeps of an incident may be required to arrive at this information. The aim of the first sweep is to extract a brief description of the incident. In the second sweep, a timeline of the main events is established. A third sweep is used to identify the key decision points along that timeline. The final sweep is concerned with probing means–ends relationships at each decision point or, in other words, exploring the reasons and resources that shaped experts' decisions.

Bisantz et al. (2003) describe another semistructured interview format, which they used for gathering information about a future system, specifically a naval destroyer.[*] The questions they put to experts were motivated by "abstraction hierarchy concepts" (p. 184). For example, experts were required to discuss the goals of different types of missions and how those goals are interconnected; the high-level guidelines or constraints on accomplishing

---

[*] See Figure 6.12 for the model of the naval destroyer.

those goals; the functions for fulfilling those goals and how these functions interact; the subsystems for achieving those goals and how these subsystems intersect; how the ship's ability to attain its goals would be affected if some of its subsystems were disabled; and whether other subsystems could be used as substitutes for those that were inoperative.

Like walkthroughs and talkthroughs, interviews are suitable for eliciting information from a range of experts. For instance, the unstructured format suggested by Rasmussen et al. (1994) may be employed with directors, executives, managers, or training instructors. The semistructured format based on the critical decision method may be used for conducting interviews with legislators, administrators, engineers, or maintenance personnel. In the same way, the semistructured format described by Bisantz et al. (2003) may be helpful for gathering information from strategists, front-line workers, scientists, or system developers.

Interviews are also similar to walkthroughs and talkthroughs in other ways. First, interviews are best conducted when analysts are familiar enough with the focus system to understand the information imparted by experts without requiring a great deal of explanation. In addition, interviews are useful for obtaining information about system functioning in situations that are unforeseen or unfamiliar. The semistructured format based on the critical decision method is well suited for this purpose since it is designed for examining nonroutine or challenging incidents. The unstructured and semistructured formats suggested by Rasmussen et al. (1994) and Bisantz et al. (2003) may be tailored to this objective. For example, experts may be asked to explain how the functions in which they participate would change during such an event or how a system's goals or subsystems would be affected in this type of situation. As is the case for walkthroughs and talkthroughs, experts must be in a position to establish without significant difficulty how a system would perform under those conditions.

Finally, some kinds of interviews are suitable for learning about future systems. As mentioned earlier, the semistructured format presented by Bisantz et al. (2003) was used to uncover information about a future naval destroyer. Furthermore, the unstructured format recommended by Rasmussen et al. (1994) may be modified to study the functions that are necessary in a future system, including the reasons and resources for those functions. Like for walkthroughs and talkthroughs, however, experts must have sufficient knowledge of the future system to respond to questions about the constraints on actors' behavior without undue effort or hesitation. Given that the critical decision method format relies on experts' prior experiences, this approach may be beneficial for learning about only the class of system to which a future system belongs, unless existing systems from which critical incidents are drawn are highly similar to the future system.

Tabletop Analysis

Tabletop analysis is a technique that engages a group of experts in thinking about a specific question or topic (Kirwan and Ainsworth, 1992). As this technique is designed expressly to support experts in problem solving or exploration, the topic of a tabletop session need not be well understood in advance. A key strength of this technique is its suitability for investigating complex questions.

For work domain analysis, tabletop discussions may be conducted to examine a range of subjects relating to the constraints on actors' behavior. For instance, experts may be tasked with exploring the criteria for evaluating organizational performance; the reasons and resources for a system's functions, tasks, or operating procedures; the objectives or values of the system; the technical capabilities or limitations of the equipment available to

BOX 8.6 WHAT ARE THE SOURCES OF INFORMATION FOR THE ANALYSIS?

- What documents, field observations, or subject matter experts are useful for gathering information about the focus system?

HINTS

- Unlike normative or descriptive techniques, work domain analysis makes use of the three information sources to establish not *what* actors do or should do, but *why* they must do something and the alternatives for *how* this can be done.
- The sources are rarely explicit about the constraints on actors' behavior, but they may be utilized to glean or piece together this information.
- In performing work domain analysis, I have found it efficient to produce a preliminary model of the focus system on the basis of information from documents and exploratory field observations. Focused field observations and walkthroughs, talkthroughs, interviews, or tabletop analyses with subject matter experts are useful for refining the preliminary model. I have also found that it is often possible to produce a reasonably comprehensive model based solely on information available in documents.
- When eliciting information from subject matter experts, the 'language' or terminology of the questions they are asked should be tailored to their backgrounds or experience.
- It is important to bear in mind that the sources may provide incorrect or conflicting information about the constraints on actors' behavior. This is one reason why analysts need to have considerable knowledge of a system to develop a work domain model. Work domain analysis does not involve simply transferring information from documents, field observations, or subject matter experts into a model but requires a significant degree of critical thinking.

workers; or the structural properties of the environment that shaped workers' decisions during a critical incident.

A variety of experts may participate in a single tabletop session. Consequently, this technique is valuable for bringing different perspectives to bear on the problem under discussion. In examining a critical incident, for example, frontline workers may explain their decisions in that situation, including the reasons and resources behind their decisions. In addition, managers or administrators may shed light on the impact of workers' decisions on the system's financial performance, whereas engineers or maintenance personnel may provide insight into the effects of those decisions on the system's technical operation. By taking advantage of the accounts of a range of experts, then, analysts can build a comprehensive picture of the constraints on actors' behavior.

As is the case for the preceding knowledge-elicitation techniques, conducting a tabletop analysis is most efficient when analysts have sufficient knowledge of the focus system to follow experts' discussions without continually asking them for clarification. It is also important to point out that tabletop analysis offers some key advantages over the other techniques. Specifically, it is ideal both for examining system functioning under new or unusual conditions and for studying future systems. Like for the other techniques, experts may be asked about how a system's functions and subsystems would be affected during an unforeseen event or what functions and subsystems are necessary in a future system. However, in the case of tabletop analysis, experts need not know the answers to such questions beforehand, which is significant because the questions can be quite demanding. Instead, given the problem-solving or exploratory format of tabletop analysis, experts can systematically investigate the answers to such questions on the basis of their collective knowledge or experience (Naikar et al., 2003, 2006).

Theme 7: What Is the Content of the Abstraction–Decomposition Space?

Central to the process of performing a work domain analysis is the question of the content of the abstraction–decomposition space. In other words, what levels of abstraction and decomposition and what constraints should comprise a model of the focus system? By and large, these are regarded as the core concerns of this form of analysis.

As explained in Chapter 2, the abstraction–decomposition space models a work domain—or the functional structure of the environment of actors—in the form of a matrix.* The vertical and horizontal axes of the matrix are delineated by a set of levels of abstraction and decomposition, respectively. The abstraction dimension, which is made up of a series of qualitatively distinct concepts for modeling the structural properties of the environment, is characterized by structural means–ends relations. The decomposition dimension, which includes a number of levels of detail or resolution for modeling the structural properties of the environment, is defined by part–whole relations. The cells of this matrix, then, are populated with the structural properties of the environment at different levels of abstraction and decomposition. These properties place constraints on actors' behavior.

The following sections provide strategies for defining the levels of abstraction and decomposition and the constraints that should be represented in the model of the focus system. These considerations in producing a work domain model are dealt with here

---

BOX 8.7   WHAT IS THE CONTENT OF THE ABSTRACTION–DECOMPOSITION SPACE?

- What levels of abstraction should I include in the model of the focus system?
- What levels of decomposition should I incorporate into the model?
- What constraints should I represent in the model?

HINTS

- Defining the content of an abstraction–decomposition space is not necessarily a linear process with distinct stages, especially for experienced analysts.
- The strategies for developing an abstraction–decomposition space are also applicable for producing an abstraction hierarchy.
- See Boxes 8.7a to 8.7d for more specific hints relating to the levels of abstraction and decomposition and the constraints to represent in a work domain model.

---

* See Figure 2.5 for the abstraction–decomposition space.

one at a time for the sake of convenience, but, in practice, these processes are nonlinear and tightly intertwined. It is also important to emphasize that these strategies are useful not only for developing an abstraction–decomposition space but also for constructing an abstraction hierarchy. As the constraints in an abstraction hierarchy do rest at some level of decomposition,* these strategies can help analysts to produce, with economy, a well-formed hierarchy that encompasses the most informative representations of the focus system.

Abstraction Dimension

Building a work domain model of the focus system involves thinking about the number, types, and labels of the levels of abstraction that should be represented in the model. This section presents two approaches for defining the abstraction dimension of the model. The first is to adopt the standard set of five levels described by Rasmussen (1986) and Rasmussen et al. (1994). As observed in Chapter 3, these levels have been useful across a range of systems and applications. Therefore, it may not be necessary to analyze the levels of abstraction to include in the model "from scratch in every case" (Rasmussen et al., 1990, p. 35).

The second approach is to investigate the levels of abstraction that should be represented in the model and not assume the standard set. The fact that the standard levels have been used in a variety of models may simply reflect that analysts were working from the example established by Rasmussen (1986) and Rasmussen et al. (1994), as explained in Chapter 3. This leaves open the possibility that, in some cases at least, a different set may provide a more faithful representation of a system, which would allow analysts to create superior designs or products for workers. Certainly, Vicente and Wang's (1998) models of the games of chess† and baseball‡ appear to depart from the standard levels. These observations indicate that analyzing the levels of abstraction to include in the model of the focus system without presupposing the standard set may be desirable.

Standard Set

A straightforward approach for specifying the abstraction dimension of a work domain model is to adopt Rasmussen's (1986) and Rasmussen et al.'s (1994) standard set of five levels of abstraction. The labels ascribed to these levels, which are those most recently suggested by Rasmussen, are functional purposes, value and priority measures, purpose-related functions, object-related processes, and physical objects (Reising, 2000). Comprehensive descriptions of each level, including examples from models of a home, a library, a military system, and a pasteurization plant, are provided in Chapter 3. Both these labels and descriptions may be better suited to intentional systems than to causal systems. Rasmussen (1986) and Vicente (1999) provide an account of the standard levels that may be more appropriate for causal systems.

If analysts decide to adopt the standard set for the model of the focus system, they will not need to analyze the levels of abstraction to include in the model but will need to define the constraints to represent at each level. The process of identifying constraints is addressed in greater depth later in this chapter, but, for the sake of clarity, Figure 8.7 provides a number of prompts and keywords to help analysts with this task in relation to the standard levels. The prompts portray the kinds of questions that analysts may find useful

---

* Chapter 4 discusses the relationship between the abstraction hierarchy and decomposition.
† See Figure 3.9 for the model of chess.
‡ See Figure 3.10 for the model of baseball.

	Prompts	Keywords
Functional Purposes	• Why does the system exist or why is the system necessary? • What are the system's principal services or outputs? • What are the system's values? • What objectives is the system designed to achieve? • What constraints or values do external entities or society impose on the system? • What external constraints is the system designed to fulfil?	Reasons, purposes, goals, aims, objectives, intentions, plans, services, outputs, products, roles, values, conventions, norms, customs, principles, laws, policies, regulations, standards, procedures, guidance, requirements, rules, limits
Value and Priority Measures	• What fundamental laws, principles, or values must be respected by the system? • What criteria must be met for the purpose-related functions of the system to achieve its functional purposes? • What criteria can be used for evaluating how well the system is fulfilling its functional purposes? • What criteria can be used for comparing, prioritizing, and allocating resources to the purpose-related functions of the system? • What flows, balances, or distributions of mass, energy, money, people, or information are necessary in the system so that it achieves its functional purposes?	Criteria, measures, success, effectiveness, efficiency, reliability, quality, quantity, economy, consistency, frequency, probability, time, risks, budgets, schedules, performance, outcomes, results, targets, figures, tests, assessments, resources, laws, principles, values, conventions, norms, customs, policies, regulations, standards, procedures, guidance, requirements, rules, limits
Purpose-related Functions	• What functions must the system be capable of supporting so that it achieves its functional purposes? • What functions are afforded by the system's physical objects? • What functions are afforded by the system's object-related processes? • What uses are physical resources put to in the system? • What functions do people in the system perform?	Functions, operations, processes, activities, roles, responsibilities, jobs, tasks, duties, occupations, positions
Object-related Processes	• What functional processes or functional capabilities or limitations of physical objects are of relevance to the system? • What functional processes or functional capabilities of physical objects are necessary for the system to achieve its purpose-related functions? • What can physical objects of relevance to the system do or afford?	Processes, capabilities, limitations, functionality, characteristics, capacity, functions, applications
Physical Objects	• What physical objects are of relevance to the system? • What physical objects are necessary to achieve the system's purpose-related functions? • What physical objects are necessary to enable the system's object-related processes? • What physical objects are present in the system?	Equipment, tools, artifacts, premises, infrastructure, facilities, fixtures, people, personnel, geographical features, assets, resources

FIGURE 8.7

Prompts and keywords for defining constraints in relation to the standard levels.

to think about in specifying the constraints to depict at each level. The keywords highlight the different forms in which details about constraints may be revealed to analysts, specifically, as they examine various information sources.

For example, the prompts suggest that to define a system's functional purposes it may be helpful to reflect on why the system exists, what objectives the system is designed to achieve, or what constraints or values are imposed on the system by external entities or society. The keywords indicate that documents, field observations, or subject matter experts may disclose details about the system's functional purposes in the guise of information about goals, aims, outputs, products, conventions, or regulations. In the same way, at the value and priority measures level, the prompts highlight that it may be useful to contemplate what fundamental laws, principles, or values must be respected by the system, what criteria must be met so that the purpose-related functions achieve the system's functional purposes, or what flows, balances, or distributions of mass, energy, money, people, or information are necessary in the system. The keywords denote that the information sources may reveal details about the system's value and priority measures in the form of descriptions of effectiveness, efficiency, reliability, budgets, schedules, performance, policies, procedures, or rules.

As implied above, the prompts and keywords may be utilized with a variety of sources of information. Specifically, they may guide analysts in searching through documents or conducting field observations to uncover data about constraints. In addition, they may be used as a reference for defining suitable topics or questions for knowledge-elicitation sessions with subject matter experts.

It is important to remember that the material uncovered by the prompts and keywords may not always signify constraints. For example, some of the uses that physical resources are put to in a system (see the prompts at the purpose-related functions level) may not be functions that the system must be capable of supporting to achieve its functional purposes or, in other words, may not be purpose-related functions. Instead, the prompts and keywords highlight information that analysts may find useful to *consider* in defining constraints. As mentioned before, the process of performing a work domain analysis does not involve simply shifting information from documents, field observations, or subject matter experts into a model but requires a significant degree of critical reasoning.

The prompts and keywords are utilized most effectively in combination with the detailed descriptions of the standard levels in Chapter 3. These descriptions impart supplementary information about each level, such as that purpose-related functions are independent of the properties of physical objects, whereas object-related processes are not. In addition, the descriptions provide clarification. As a case in point, they shed light on the fact that the duplication of some keywords at the functional purposes and value and priority measures levels in Figure 8.7 indicates that a system's values may be apparent at both levels, albeit in the form of the objectives to be attained at the first level and the criteria to be respected at the second level. The descriptions also highlight variations that analysts may adopt in modeling constraints. For instance, a system's secondary objectives or external constraints may not be represented explicitly at the functional purposes level but may be portrayed implicitly in the form of criteria at the value and priority measures level.

Finally, one or more of the standard levels may be excluded from the model of the focus system, either because they are not suited to the characteristics of the system or because they are not necessary for the model's intended application. For example, as discussed in Chapter 3, some models of computer simulations do not depict the lowest level of the standard set because those simulations do not have a physical form (Burns, 2000; Jamieson and Vicente, 2001; Reising and Sanderson, 2002b). Similarly, Burns and Vicente (1995, 2000)

BOX 8.7A   HINTS FOR STANDARD SET OF LEVELS OF ABSTRACTION

- Figure 8.7 provides prompts and keywords for defining the constraints to represent at each of the standard levels in a model. The prompts suggest questions for analysts to contemplate in identifying constraints, whereas the keywords indicate the guises in which constraints may be revealed.
- The prompts and keywords may be utilized with a variety of information sources.
- The prompts and keywords may not always uncover material that signifies constraints, but highlight information that analysts find useful to consider in defining constraints.
- The prompts and keywords are best employed in conjunction with the descriptions of the standard levels in Chapter 3. Specifically, the descriptions present supplementary information, provide clarification, and highlight variations that analysts may adopt in modeling constraints.
- Analysts may exclude one or more of the standard levels from the model of the focus system, depending on the characteristics of the system or the requirements of an application.

included only three of the five standard levels in their model of an engineering design system because these levels were sufficient for addressing their design goals.

Indeterminate Set

A more demanding approach for defining the abstraction dimension of a work domain model is to establish the levels of abstraction to include in the model without presupposing the standard set. The strategy proposed here for this purpose recognizes that this dimension is distinguished by several distinct concepts for modeling the functional structure of the environment, and that these concepts are linked by structural means–ends relations. Therefore, by identifying the structural properties of the environment of actors, classifying these properties into distinct concepts, and organizing these concepts into a hierarchy of structural means–ends relations, analysts can specify the levels of abstraction to include in the model so that it offers a faithful characterization of the focus system.

To assist analysts with this strategy, Table 8.1 provides a number of prompts and keywords. The prompts, which highlight the kinds of questions that analysts may reflect on

TABLE 8.1

Prompts and Keywords for Uncovering the Structural Properties of the Environment

Prompts	Keywords
• What are the properties of the physical, social, or cultural environment of actors?	Properties of physical, social, or cultural environment, reasons, resources, constraints, limits, affordances, capabilities
• What are the reasons for actors' behavior?	
• What are the resources for actors' behavior?	
• What are the constraints or limits on action, irrespective of the situation?	
• What are the affordances or capabilities for action, irrespective of the situation?	

to identify the structural properties of the environment, recognize that these properties originate in the physical, social, or cultural context of actors. Such properties define the fundamental reasons and resources for actors' behavior. Furthermore, they not only place limits on action but also provide capabilities for action, both of which are event independent. The keywords emphasize the main idea behind each prompt, which analysts may find convenient to keep in mind while examining the information sources for relevant material. Like the previous prompts and keywords, those in Table 8.1 may be better suited to intentional systems than to causal systems, and they may be utilized with a variety of sources of information.

The preceding strategy may be illustrated with reference to the model of a home (see Figure 2.7). As explained before, the purpose of this model was to provide examples for the material in this book that originate from a system that is not only familiar to readers but also intentional rather than causal in nature. In this case, the boundaries of the analysis may focus on the organizational entity of the home, the physical entity of the house, and the problem of residing in the home from the perspective of actors who live there.

Using the prompts and keywords in Table 8.1 to define the structural properties of the environment of actors in this focus system, it may be established that *prepare meals and beverages*, *cooling*, and *heating* are capabilities for action in the home, whereas *time* and *money* place constraints or limits on actors' behavior (Figure 8.8a). It may also be determined that a *refrigerator* and *stove* are resources for actors' behavior in the home, while *well-being* is a reason for their actions.

When classifying these properties into distinct concepts, *prepare meals and beverages* may be identified as representing a fundamentally different kind of concept from *time* because the former signifies a function that must be available in the home, whereas the

Structural properties of the environment of actors	Concepts	Hierarchy of structural means-ends relations
Prepare meals and beverages Cooling Heating Time Money Refrigerator Stove Well-being	<u>Functions</u> Prepare meals and beverages <u>Criteria</u> Time Money <u>Physical Objects</u> Refrigerator Stove <u>Functional Capabilities of Physical Objects</u> Cooling Heating <u>Objectives</u> Well-being	<u>Objectives</u> Well-being <u>Criteria</u> Time Money <u>Functions</u> Prepare meals and beverages <u>Functional Capabilities of Physical Objects</u> Cooling Heating <u>Physical Objects</u> Refrigerator Stove
(a)	(b)	(c)

FIGURE 8.8
Illustration of a strategy for analyzing the levels of abstraction to include in a model without presupposing the standard set: (a) identifying the structural properties of the environment; (b) classifying the properties into distinct concepts; (c) organizing the concepts into a hierarchy of structural means–ends relations.

latter denotes a criterion that must be respected by actors in this system (Figure 8.8b). In addition, *time* and *money* are similar concepts as they both signify criteria. In the same way, it may be established that *refrigerator* and *stove* are comparable concepts because they are both physical objects, and that *cooling* and *heating* are alike given that they both denote functional capabilities of physical objects. Last, it may be decided that *well-being* is an objective of the home, and that this concept is fundamentally different from any of the others that have been identified.

When organizing these concepts into a hierarchy of structural means–ends relations, it may be ascertained that the objective of *well-being* is an end that can be achieved by criteria relating to *time* and *money* (Figure 8.8c). Likewise, these criteria are ends that can be met by functions such as *prepare meals and beverages*. This function, in turn, is an end that can be attained by the functional capabilities of physical objects such as *cooling* and *heating*. These functional capabilities can be realized by physical objects such as a *refrigerator* and a *stove*. In this example, then, a set of five levels of abstraction has been identified for the model of the focus system. Figure 8.9 provides another illustration of this strategy.

Several points are worth bearing in mind with respect to the levels of abstraction that are included in a model of the focus system. First, these levels should provide qualitatively different views of the system, rather than just add or subtract details from the same view.[*] In addition, the relationships between the levels should be structural means–ends relations.[†] Furthermore, if one or more of the levels appear to deviate from the standard set, it is important to check that these levels signify concepts that are fundamentally different from the standard levels and not the same concepts at different levels of detail. For example, analysts may classify the structural properties of the environment into purpose-related functions at two different levels of detail (or decomposition), and mistakenly view one of these levels as unique. It is also worth checking that these levels denote concepts that relate to the structural properties of the environment and not other aspects of the system, such as its activities.

A useful approach for implementing the preceding strategy when the project restrictions are tight is to limit the analysis of the abstraction dimension to essential or critical information sources, which may be just one or more documents, sets of field observations, or subject matter experts. The prompts and keywords in Table 8.1 may be used to examine these critical sources for the structural properties of the environment. At this stage, these properties only need to be recorded in a form that is suitable for defining the concepts, or levels of abstraction, to include in the model (e.g., *money*), not necessarily in the form of constraints (e.g., *total income – total expenses = savings*). Similarly, analysts need not be overly concerned about whether such properties are noted in a form that is event independent (e.g., *preparation of meals and beverages*) or in a form that emphasizes behavior (e.g., *prepare meals and beverages*), which is event dependent, provided that the structural properties can be easily inferred in the case of the latter.

Subsequently, the list of properties can be examined to determine whether or not they correspond to one of the levels in the standard set.[‡] At this point, it is not necessary to specify exactly to which of the standard levels, if any, each of the properties relates. For example, analysts need not be sure whether properties such as *cooling* and *heating* signify object-related processes or purpose-related functions of the home. Instead, it is sufficient

---

[*] Chapter 2 discusses the fact that the levels of abstraction in a model represent distinct concepts.

[†] Structural means–ends relations are explained in Chapters 2 and 5.

[‡] See Chapter 3 for comprehensive descriptions of the standard levels and the preceding section of this chapter for prompts and keywords relating to this set.

BOX 8.7B   HINTS FOR INDETERMINATE SET
OF LEVELS OF ABSTRACTION

- The abstraction dimension may be specified by identifying the structural properties of the environment of actors, classifying these properties into distinct concepts, and organizing these concepts into a hierarchy of structural means–ends relations. This strategy is not necessarily a linear process with discrete steps, particularly for experienced analysts.
- The prompts and keywords in Table 8.1 may be helpful for uncovering the structural properties of the environment. These prompts and keywords may not always lead directly to such properties, but they highlight information that analysts may find useful to *consider* in identifying the properties.
- The structural properties of the environment must fall within the boundaries of the analysis or, in other words, must be relevant to the focus system.
- It may be beneficial to keep more extensive records of the structural properties of the environment than are shown in Figures 8.8a and 8.9a. For instance, the description of each property may comprise several sentences from documents, notes of field observations, or transcripts of knowledge-elicitation sessions with subject matter experts. Details about the sources of this information (e.g., document names, page numbers) may also be worth recording in case it is necessary to return to it later. In particular, this information may be useful for classifying the properties into distinct concepts, organizing the concepts into a hierarchy of structural means–ends relations, and defining constraints.
- The levels of abstraction in the model of the focus system should offer distinct views of the system rather than just add or remove details from the same view (see Chapter 2). In addition, the relationships between these levels should be structural means–ends relations (see Chapters 2 and 5).
- If one or more levels of abstraction in the model appear to depart from the standard set, it is worth checking that they represent concepts that are qualitatively different from those of the standard levels instead of the same concepts at different levels of detail. It is also worth ensuring that they represent concepts that relate to the structural properties of the environment and not to other aspects of the system.
- An efficient process for defining the abstraction dimension of a model is to examine critical information sources for the structural properties of the environment and establish whether these properties are the same as or different from those in the standard set. If these properties and those in the standard set are the same, the standard set may be adopted for the model of the focus system. If they are different, the abstraction dimension may be specified by sorting the properties into distinct concepts and arranging these concepts into a hierarchy of structural means–ends relations.
- Analysts may be unable to define the abstraction dimension of the model fully until they have addressed, at least partially, other aspects of the model, such as its levels of decomposition and its constraints.
- Analysts may exclude some levels of abstraction from the model of the focus system if these levels are unnecessary for the intended application.

Structural properties of the environment of actors	Concepts	Hierarchy of structural means–ends relations
Bathing Showering Conservation of natural resources Protect the environment Personal care and grooming Bath Shower	<u>Functional Capabilities of Physical Objects</u> Bathing Showering <u>Criteria</u> Conservation of natural resources <u>Objectives</u> Protect the environment <u>Functions</u> Personal care and grooming <u>Physical Objects</u> Bath Shower	<u>Objectives</u> Protect the environment <u>Criteria</u> Conservation of natural resources <u>Functions</u> Personal care and grooming <u>Functional Capabilities of Physical Objects</u> Bathing Showering <u>Physical Objects</u> Bath Shower
(a)	(b)	(c)

FIGURE 8.9
Second illustration of a strategy for specifying the abstraction dimension of a model without presuming the standard set: (a) uncovering the structural properties of the environment; (b) sorting the properties into distinct concepts; (c) arranging the concepts into a hierarchy of structural means–ends relations.

for analysts simply to establish whether or not a property relates to one or another of the standard levels.

If all of the properties evident in critical information sources signify concepts that are the same as those in the standard set, analysts may choose to adopt the standard levels for the model of the focus system. Otherwise, analysts may define the levels of abstraction to portray in the model by classifying the properties into distinct concepts and organizing these concepts into a hierarchy of structural means–ends relations.

It is worth mentioning here that my colleagues and I have implemented this process, with slight variations, on three projects involving the analysis of a maritime surveillance system (Elix and Naikar, 2008b), an air power system (Lambeth and Naikar, 2008), and an air combat system (Treadwell and Naikar, 2009). Furthermore, alternative processes more closely resembling the strategy illustrated in Figures 8.8 and 8.9 were adopted for the analysis of an advanced fighter aircraft (Hopcroft and Naikar, 2005), a manned space system (Baker, 2006; Baker et al., 2008), and a commercial training helicopter (Baker, 2006). In no case did we find evidence for a set of levels of abstraction that diverged from the standard set, which suggests that the standard levels provide a valid characterization of these systems. However, it is also possible that the processes we implemented are incapable of uncovering a different set. Therefore, it may be fruitful to direct future research at developing suitable processes or strategies for this purpose.

Decomposition Dimension

Developing a work domain model of the focus system also involves thinking about the number, types, and labels of the levels of decomposition to include in the model. In contrast to the situation for the abstraction dimension, analysts have not adopted a standard set of levels of decomposition for modeling a variety of systems, as discussed in Chapter 4. Instead, the decomposition dimension across a range of models appears to vary with the characteristics of a system or the requirements of an application. One apparent pattern is that the majority of models are characterized by physical decomposition, which means that the levels are based on functional aggregations of physical objects. A few models are defined by conceptual decomposition, such that the levels are based on functional aggregations of an assortment of concepts, possibly including physical objects.[*] However, there are insufficient cases to conclude with certainty that all systems can be modeled faithfully with either physical or conceptual decomposition. These observations suggest that a strategy for defining the decomposition dimension of a model in the absence of any preconceptions about its characteristics is necessary.

Before presenting such a strategy, it is important to note that, regardless of the number, types, and labels of the levels of decomposition in a model, there are two approaches that analysts have employed to create representations of a system at different levels of detail. In the case of constraint decomposition, the constraints in a model are decomposed or aggregated into respectively more or less detailed representations of a work domain.[†] System decomposition, on the other hand, involves producing a decomposition hierarchy by decomposing a system into parts or aggregating the parts of a system into wholes. Then, using either full or partial abstraction hierarchies, progressively more detailed work domain representations are constructed for the parts at each level of decomposition as one proceeds along this dimension from left to right.[‡]

The strategy proposed here for defining the decomposition dimension of a model presumes system decomposition, which is by far the more common approach. As discussed in Chapter 4, this approach allows analysts to assemble a skeletal abstraction–decomposition space of the focus system, thereby gaining an overview of the potential content of the cells in the model. This overview makes it possible not only to make informed decisions about which representations of a work domain are most worth developing in view of the purpose of the analysis but also to craft models with higher levels of comprehensiveness, accuracy, and efficiency. These benefits of a skeletal abstraction–decomposition space are also relevant when analysts intend to construct an abstraction hierarchy rather than an abstraction–decomposition space. In this case, it may be prudent to formulate a skeletal abstraction–decomposition space of the focus system by studying, at a minimum, just a few of its parts in order to specify the decomposition dimension of the model.

As pointed out in Chapter 4, system decomposition has two variations. Both variations entail building a decomposition hierarchy by decomposing a system into parts or aggregating the parts of the system into wholes. In the first variation, full or partial abstraction hierarchies are then formulated for all or many parts of the system at each level of decomposition, so that comprehensive work domain representations encompassing a range of parts are produced at each level.[§] In the second variation, full or partial abstraction hierarchies are composed for only a single part of the system at each level of decomposition, resulting in a detailed work domain representation of a specific part at each level.[¶] The strategy presented next focuses on the process of generating a decomposition hierarchy. Whether analysts subsequently choose to adopt the first or second variation of system decomposition depends on the requirements of the intended application.

---

[*] Chapter 4 discusses physical and conceptual decomposition in more detail.
[†] See Figure 4.3 for an illustration of constraint decomposition.
[‡] See Figures 4.6a and 4.6b for illustrations of system decomposition.
[§] See Figure 4.6a for an illustration of the first variation of system decomposition.
[¶] See Figure 4.6b for an illustration of the second variation of system decomposition.

The proposed strategy recognizes that the decomposition dimension constitutes a number of levels of detail or resolution for modeling the structural properties of the environment. These levels of resolution are defined by the functional parts of a system, which actors need to reason about, or act on, for the system to perform effectively. The relationships between these levels are part–whole relations. Consequently, by identifying the functional parts of the focus system, classifying these parts into different levels of resolution, and organizing these levels of resolution into a hierarchy of part–whole relations, it becomes possible to specify the decomposition dimension of the model such that it offers a valid characterization of the system.

Table 8.2 provides a number of prompts and keywords to help analysts with the task of identifying a system's functional parts. The prompts indicate questions that may be useful to consider in defining such parts, for instance, regarding the coarsest part of the system about which actors need to reason or the different parts of the system around which work is organized. The keywords provide clues to the types of details that may be worth investigating further to uncover a system's functional parts, such as details about its subsystems, physical arrangements, or roles. Like the prompts and keywords presented earlier, those in Table 8.2 may be used with a variety of information sources.

As for the abstraction dimension, the strategy for defining the decomposition dimension of a model may be illustrated with a work domain analysis of a home, which was conducted with the goal of providing examples for the ideas in this book (see Figure 2.7 for the resulting model). The focus system for this analysis was outlined in the preceding section.

Using the prompts and keywords in Table 8.2 to define the functional parts of this focus system, it may be established that the coarsest part of the home that actors need to reason about, or act on, is the entire *house* (Figure 8.10a). On the other hand, some of the most fine-grained parts of this system are an *armchair* and a *toothbrush*. It may also be identified that some of the different parts of the home around which work is organized are the *lounge room*, *bathroom*, *coffee table*, and *bath*.

When classifying the functional parts of this system into different levels of resolution, it may be determined that the *house* is at a separate level from the *armchair*, as the *house* constitutes the entire physical composition of the home, whereas the *armchair* is one of its

TABLE 8.2

Prompts and Keywords for Identifying the Functional Parts of a System

Prompts	Keywords
• What is the coarsest or largest part of the system that actors need to reason about or act on?	Systems, subsystems, components, physical structures, physical spaces, physical arrangements, organizations, departments, groups, functions, positions, roles
• What are the most fine-grained or smallest parts of the system that actors need to reason about or act on?	
• What are the different parts of the system that actors need to reason about or act on?	
• What is the coarsest or largest part of the system around which work or action is organized?	
• What are the most fine-grained or smallest parts of the system around which work or action is organized?	
• What are the different parts of the system around which work or action is organized?	

Functional parts	Levels of resolution	Part-whole relations between levels of resolution
House Lounge room Bathroom Coffee table Bath Armchair Toothbrush	<u>House</u> <u>Contents</u> Armchair Toothbrush Coffee table Bath <u>Rooms or Subspaces</u> Lounge room Bathroom	<u>House</u> <u>Rooms or Subspaces</u> Lounge room Bathroom <u>Contents</u> Armchair Toothbrush Coffee table Bath

| (a) | (b) | (c) |

FIGURE 8.10

Illustration of a strategy for analyzing the levels of decomposition to include in a model without assuming a prespecified set: (a) identifying the functional parts of a system; (b) classifying the parts into different levels of detail or resolution; (c) organizing the levels of resolution into a hierarchy of part–whole relations.

contents (Figure 8.10b). In contrast, items such as the *armchair, toothbrush, coffee table,* and *bath* all belong to the same level since they are all contents. It may also be established that the *lounge room* is at a separate level from either the house or the contents as the *lounge room* is one of the rooms or subspaces of the home. Last, the *lounge room* and the *bathroom* belong to the same level because they are both rooms or subspaces.

When organizing the levels of resolution into a hierarchy of part–whole relations, it may be established that the house is composed of rooms or subspaces, such as the *lounge room* and the *bathroom* (Figure 8.10c). The rooms or subspaces, in turn, are composed of contents. Specifically, the *armchair* and the *coffee table* are functional parts of the *lounge room*, whereas the *toothbrush* and the *bath* are functional parts of the *bathroom* (Figure 8.11). Hence, in this case, a set of three levels of decomposition has been specified for the model of the focus system. The decomposition hierarchy may be developed to a level of comprehensiveness that is suitable for the purpose of the analysis.

It is important to point out that, for any focus system, there may be a number of alternative functional aggregations that could be adopted for the model. For example, it may be possible to group the parts of the home into functional aggregations other than whole house, rooms or subspaces, and contents. The particular groupings that are selected for the model will depend on the purpose of the analysis. As Vicente (1999) advises, the parts of a system within a functional aggregation should be tightly coupled to each other but weakly coupled to the parts in other functional aggregations. In relation to the model of a home, for instance, the parts or contents of the bathroom are tightly coupled to each other but weakly coupled to the contents of the other rooms or subspaces.

Another check worth performing on the levels of decomposition that are included in the model of the focus system is that each level offers a different resolution for viewing a work domain.[*] More specifically, higher levels should provide a coarse resolution, whereas

---

[*] Chapter 2 explains that the levels of decomposition in a model offer different resolutions for viewing a work domain.

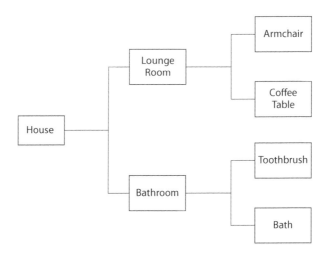

FIGURE 8.11
Illustration of the process of organizing the functional parts at each level of resolution into a hierarchy of part–whole relations.

lower levels should provide a fine-grained resolution. It is also advisable to check that the functional parts at each level are at the same resolution and that the relationships between the levels are part–whole relations.[*]

Again, when the project restrictions are tight, it may be sensible to focus on critical documents, sets of field observations, or subject matter experts. These information sources can be examined to identify a system's functional parts using the prompts and keywords in Table 8.2. Following that, the functional parts may be sorted into distinct levels of resolution, and the levels of resolution may be arranged into a hierarchy of part–whole relations. Then, having specified a set of levels of decomposition to include in the model of the focus system, the decomposition hierarchy may be expanded simply by exploiting part–whole relations to identify the additional functional parts at each level. For example, with respect to the decomposition hierarchy of a home (Figure 8.11), analysts may seek to define all of the rooms or subspaces of the *house* or all of the contents of the *bathroom*. As noted previously, the decomposition hierarchy may be constructed as extensively as is required for the purpose of the analysis.

My colleagues and I have applied this process, or minor variations of it, to analyze six military and aerospace systems (Baker, 2006; Baker et al., 2008; Elix and Naikar, 2008b; Hopcroft and Naikar, 2005; Lambeth and Naikar, 2008; Treadwell and Naikar, 2009). This process led to a decomposition hierarchy with a physical basis for all but one of the systems. Specifically, our analysis of the air power system found that it could be characterized by either physical or conceptual decomposition, although the former was selected because it was better suited to the intended application of the model. These findings suggest that the proposed process is capable of uncovering decomposition hierarchies with either a physical or a conceptual basis. Whether it has the capacity to reveal other types of decomposition hierarchies, if they exist, has not been established.

---

[*] See Chapters 2 and 5 for descriptions of part–whole relations.

BOX 8.7C   HINTS FOR DECOMPOSITION DIMENSION

- The majority of models, across a range of systems, are characterized by physical decomposition, whereas a few models have a decomposition dimension with a conceptual basis.
- The decomposition dimension of a model may be specified by identifying the functional parts of the focus system, classifying these parts into different levels of resolution, and organizing these levels of resolution into a hierarchy of part–whole relations. In practice, this strategy is unlikely to be executed as a linear process with discrete steps, particularly by experienced analysts.
- Table 8.2 provides a number of prompts and keywords for uncovering a system's functional parts. Although these prompts and keywords may not point directly to such parts, they draw attention to details that are worth considering in defining them.
- Actors may reason about, or act on, the functional parts of the focus system at a variety of levels of abstraction. For example, actors may reason about the rooms or subspaces of a home at the level of their functional purposes, value and priority measures, purpose-related functions, object-related processes, or physical objects (see Figure 2.7). That said, a number of studies show that actors mainly adopt models along the diagonal of the abstraction–decomposition space (see Chapter 2).
- The functional parts used for defining the decomposition dimension of a model must fall within the boundaries of the analysis. That is, they must be applicable to the focus system.
- Paralleling the case for the abstraction dimension (see Box 8.7b), keeping more detailed records of a system's functional parts than are presented in Figure 8.10a may be worthwhile. These details may be useful for categorizing the functional parts into different levels of resolution, ordering the levels of resolution into a hierarchy of part–whole relations, and defining constraints.
- The decomposition hierarchy may be developed as comprehensively as is necessary for the purpose of the analysis.
- It may be possible to group the parts of the focus system into a number of alternative functional aggregations. The groupings that are adopted will depend on the purpose of the analysis. Note that parts within a functional aggregation should be strongly coupled to each other but weakly coupled to those in other functional aggregations.
- It is also worth checking that the levels of decomposition in the model offer different resolutions for viewing the work domain, with higher levels providing coarse resolutions and lower levels providing fine-grained resolutions (see Chapter 2). In addition, the functional parts at each level should denote the same resolution, and the relationships between the levels should be part–whole relations (see Chapters 2 and 5).
- An efficient process for defining the decomposition dimension of a model is to focus on critical information sources in order to formulate the initial set of levels of decomposition. Following that, the decomposition hierarchy may be extended principally by taking advantage of part–whole relations to identify further functional parts at each level.

- An apparent duplication may be evident between the abstraction and decomposition hierarchies of a model when the decomposition dimension has a physical basis, or a conceptual basis with one or more levels defined by physical objects, and the names of physical objects are listed at one of the levels of abstraction (see Chapter 5).
- It may not be possible to define the decomposition dimension of a model completely until other aspects of the model, such as its levels of abstraction and its constraints, have been established at least partially.
- Some levels of decomposition may be excluded from the model if these are not relevant for the purpose of the analysis.

Constraints

The process of constructing a work domain model also requires analysts to think about the constraints that should be included in the model. This section addresses four questions that are worth contemplating when defining constraints: What is the potential content of the cells in the abstraction–decomposition space? Which parts of the system at each level of decomposition should be modeled with constraints? Which cells of the model should be populated with constraints? and What constraints should be represented in the selected cells? As was the case for the abstraction and decomposition dimensions, the strategy for specifying constraints may be explained in the context of a work domain analysis of a home (see Figure 2.7). The focus system for this analysis was described previously.

What Is the Potential Content of the Cells in the Model?

One question to take into account when defining the constraints to include in the model of the focus system is, What is the potential content of the cells in the model? As explained in Chapter 4, with this information it is possible to take a considered approach in deciding which representations of the work domain are the most useful to develop, so that the model is ideal for its intended application. It is also possible to produce, more efficiently, models of higher quality in terms of their comprehensiveness and accuracy.

An overview of the potential content of the cells in the model of the focus system can be obtained by constructing a skeletal abstraction–decomposition space. This process involves juxtaposing the abstraction and decomposition dimensions of the model to form a matrix, so that the potential content of each cell, as a function of its levels of abstraction and decomposition, is revealed. For an example, see the skeletal abstraction–decomposition space of a home presented earlier in Figure 4.12.

If analysts are leaning toward modeling the focus system with an abstraction hierarchy, they may find it beneficial to construct a skeletal abstraction–decomposition space by adopting a simplified approach. That is, they may study just a few parts of the system to define the decomposition dimension of the model, as described in Chapter 4. Even with this rudimentary approach to acquiring an overview of the potential content of the cells in the model, the resulting abstraction hierarchy is more likely to be assembled efficiently and to be of a higher quality. It is also more likely to be well suited to the proposed application.

Which Parts of the System to Model?

Another consideration when specifying the constraints to include in the model of the focus system is, 'Which parts of the system at each level of decomposition should be represented with constraints?' For example, depending on the purpose of the analysis, the first variation of system decomposition may be adopted so that all or many parts of the system at each level are considered.[*] This option produces a comprehensive work domain representation incorporating several parts at each level. Alternatively, the second variation of system decomposition may be selected, which means that a single part of the system at each level is the target of the analysis.[†] In this case, a detailed work domain representation of the nominated part is obtained at each level. The two options may be implemented within the same model but at different levels of decomposition.

When deciding which parts of the focus system to represent with constraints, a skeletal abstraction–decomposition space, encompassing the decomposition hierarchy, is helpful. This representation can be used to review systematically which parts of the system are the most beneficial to model with constraints in light of the purpose of the analysis. For an illustration of how this decision affects the nature of the model that is created, including the constraints that are represented in the model, see the skeletal abstraction–decomposition space of the home in Figure 4.13.

Which Cells of the Model to Populate?

Also important when defining the constraints to portray in the model of the focus system is the question of which cells of the model should be populated with constraints. It is worth noting that, in making this decision, analysts are effectively choosing whether to create full or partial abstraction hierarchies for the nominated parts at each level of decomposition.[‡] This decision is significant because, despite the fact that it is possible to construct work domain representations for every cell in a model, this is unlikely to be a productive or efficient path along which to proceed, as highlighted in Chapters 2 and 4. Instead, analysts should select which cells to develop by assessing the value of the information each cell offers for the purpose of the analysis. A skeletal abstraction–decomposition space of the focus system, highlighting the potential content of the cells in the model, is useful for making such judgments.

Initially, it is sensible to evaluate which cells around the diagonal of the model should be populated with constraints. These cells normally offer useful views of the work domain irrespective of the purpose of the analysis because coarse levels of resolution are more meaningful for reasoning about a system's overall purposes, whereas fine levels of resolution are more effective for considering a system's individual physical objects. Subsequently, analysts may investigate which of the remaining cells offer novel information, relative to those along the diagonal, that is pertinent to the proposed application. Work domain representations may also be formulated for these cells. To appreciate how this decision influences the nature of the resulting model, especially the constraints that are depicted in the model, see the skeletal abstraction–decomposition space of the home in Figure 4.14.

---

[*] See Figure 4.6a for the first variation of system decomposition.
[†] See Figure 4.6b for the second variation of system decomposition.
[‡] Figures 4.6a and 4.6b illustrate the development of full or partial abstraction hierarchies for the nominated parts in the model.

What Constraints to Represent in the Selected Cells?

The question of exactly what constraints should be represented in the model of the focus system must be answered in view of the levels of abstraction and decomposition of the selected cells and the nominated parts at each level of decomposition. For instance, in the case of the skeletal abstraction–decomposition space of a home (see Figure 4.14), the constraints that should be portrayed in the model, given the levels of abstraction and decomposition of the shaded cells, include the home's functional purposes and value and priority measures in relation to the entire house, purpose-related functions with respect to its rooms or subspaces, and object-related processes and physical objects with reference to its contents. More specifically, the constraints in each cell should encompass the nominated parts of the home. For some examples of constraints that belong in the selected cells of this model, see Figure 2.7.

It is important to remember that, in the context of work domain analysis, constraints are structural properties of the physical, social, or cultural environment of actors in the focus system, which place limits on their behavior.[*] This means that the process of defining constraints essentially involves identifying the structural properties of the environment that should be incorporated into the model. As stated previously, the constraints that are depicted in the model depend on the levels of abstraction and decomposition of the selected cells, as well as the nominated parts of the system. Hence, during the process of specifying constraints, a skeletal abstraction–decomposition space of the focus system is valuable for ensuring comprehensiveness, accuracy, and efficiency, as demonstrated in Chapter 4.

To help analysts with the task of defining constraints, three specific suggestions are offered next. Some examples from the home model are provided to illustrate how these suggestions may be applied to cells at different levels of abstraction and decomposition, while taking into account the nominated parts of the system.

The first suggestion is relevant when the levels of abstraction in the model of the focus system are the same as those in the standard set (Rasmussen, 1986; Rasmussen et al., 1994). In this case, the prompts and keywords in Figure 8.7 and the detailed descriptions of the standard levels in Chapter 3 may be useful for specifying constraints. For example, one of the prompts at the functional purposes level is 'Why does the system exist or why is the system necessary?' When applying this prompt to the top left cell in the skeletal abstraction–decomposition space of a home (see Figure 4.14), it may be identified that *well-being* is a reason for which the entire house exists. Accordingly, *well-being* is a functional purpose of the home at the whole house level of decomposition (see Figure 2.7). Similarly, one of the prompts at the purpose-related functions level is 'What functions must the system be capable of supporting so that it achieves its functional purposes?' In relation to the cell at the rooms or subspaces level of decomposition in the skeletal model of a home, it may be established that *preparation of meals and beverages* is a function that the kitchen, one of the nominated parts, must sustain for the home to fulfill its functional purposes. Thus, *preparation of meals and beverages* is a constraint that should be represented in this cell of the model.

Regardless of whether the levels of abstraction in the model of the focus system are the same as or different from those in the standard set, the prompts and keywords in Table 8.1, which target the structural properties of the environment, may be used for defining constraints. For instance, one of the prompts is 'What are the reasons for actors' behavior?'

---

[*] See Chapter 2 for an explanation of the types of constraints that are the focus of work domain analysis.

With respect to the skeletal model of a home, it may be determined that *total income – total expenses = savings* is a criterion for actors' behavior, and that this criterion is not specific to any of the rooms or subspaces or contents of the home but is relevant to the entire system. Hence, this constraint is a value and priority measure at the whole house level of decomposition. Another prompt in Table 8.1 is 'What are the affordances or capabilities for action, irrespective of the situation?' When applied to the skeletal model of a home, it may be ascertained that *cooling capacity* and *heating capacity* are affordances or capabilities for action that are necessary in the home, and that these properties are functional capabilities of a refrigerator and stove, which are contents of the home. Consequently, these constraints are object-related processes at the contents level of decomposition.

Last, if the abstraction dimension of the model is defined without presuming the standard levels, the structural properties of the environment uncovered during this process may be used for specifying constraints. Consider the structural properties that were noted earlier in relation to the home model (Figures 8.8a and 8.9a). It may be identified that a *refrigerator* and *stove* are physical objects that are contents of the home. Thus, these constraints should be represented in the bottom right cell in the model of a home. In the same way, it may be determined that *protect the environment* signifies an objective of the home that the entire house is designed to achieve. Therefore, *environmental protection* is a functional purpose at the whole house level of decomposition.

It is important to emphasize that, irrespective of which of the preceding suggestions are implemented, structural means–ends relations also provide a basis for defining constraints.[*] For example, in the case of the home model, constraints may be identified by examining what value and priority measures are means for fulfilling the functional purposes of *well-being* or *environmental protection*, what purpose-related functions can be achieved by the object-related processes of *cooling capacity* or *heating capacity*, or what physical objects are means for attaining the object-related processes of *cooling capacity* or *heating capacity*. Such an approach is possible once analysts have specified one or more constraints of relevance to the model of the focus system. Bear in mind that this process may involve crossing levels of decomposition. For instance, with respect to the home model, it is necessary to establish what value and priority measures at the whole house level of decomposition can be satisfied by the purpose-related functions at the rooms or subspaces level.

In specifying the constraints to include in the model of the focus system, it may be useful to develop a glossary to supplement the model. As highlighted in Chapter 7, a glossary makes it possible to describe the constraints in greater depth than the graphical or tabular format of the model allows. Among other things, a glossary may explain the meaning of the constraints in the model, the rationale for their inclusion, and the basis of the structural means–ends relations between particular constraints. For instance, a glossary for the model of a home may record that *time* is a value and priority measure because it is a criterion for actors' behavior. Specifically, it may note that actors must balance the available *time* across work, rest, and leisure in order to achieve *well-being*, and that as this criterion holds irrespective of the specific rooms or subspaces or contents of the home, it belongs in the cell at the whole house level of decomposition. Furthermore, the glossary may clarify that this criterion can be satisfied by means of the purpose-related functions since, by providing the capabilities for those functions within a single area or location, the home makes it possible for actors to balance *time* across work, rest, and leisure, and thus achieve *well-being*.

---

[*] See Chapters 2 and 5 for descriptions of structural means–ends relations.

Finally, a number of checks are worth undertaking of the constraints that are included in the model of the focus system.* These are as follows:

1. Constraints in cells at the same level of abstraction should reflect the same concepts in a common "modeling language" (Vicente, 1999, p. 171). In contrast, constraints in cells at different levels of abstraction should manifest different concepts in distinct modeling languages.

2. Constraints in cells at lower levels of abstraction should represent structural means for achieving constraints in cells at higher levels. On the other hand, constraints in cells at higher levels of abstraction should denote structural ends that can be attained by constraints in cells at lower levels.

3. Constraints in cells at the same level of decomposition should represent the work domain at the same resolution. Conversely, constraints in cells at different levels of decomposition should represent the work domain at different resolutions.

4. Constraints in cells at higher levels of decomposition should offer coarse representations of the work domain. Alternatively, constraints in cells at lower levels of decomposition should offer fine-grained representations.

Additional Guidance and Examples

When defining the constraints to represent in the model of the focus system, it may be easier to begin at particular levels of abstraction. With reference to the standard set (Rasmussen, 1986; Rasmussen et al., 1994), Vicente (1999) suggests starting at the topmost and bottom two levels. Burns and Hajdukiewicz (2004) recommend beginning at the highest level, as the system's functional purposes will constrain the rest of the model.

At any particular level of abstraction in the model, a constraint may be placed at one of several levels of decomposition. Inexperienced analysts often have difficulty establishing at which level of decomposition they should represent a specific constraint. An example may assist analysts in dealing with this problem. Consider the fact that at the purpose-related functions level in the skeletal model of a home (see Figure 4.14), a constraint may be portrayed at one of three levels of decomposition: whole house, rooms or subspaces, and contents. If the constraint is a purpose-related function of the entire house, it belongs in the cell at the whole house level of decomposition. Alternatively, if the constraint is a purpose-related function of one of the rooms or subspaces of the home, it belongs in the cell at the second level of decomposition. Note, however, that some of the other rooms or subspaces may also be means for achieving that purpose-related function. For instance, both the bedroom and the lounge room may be means for achieving the purpose-related function of *rest*. Last, if the constraint is a purpose-related function of one of the contents of the home, it belongs in the cell at the last level of decomposition. As before, some of the other contents may also provide means for attaining that purpose-related function.

The constraints in the model may be specified as comprehensively as is required for the purpose of the analysis. As a result, some aspects of the model may be developed more fully than others. If analysts are unsure about the requirements of the proposed application, it may be efficient to develop the model broadly and, subsequently, add detail as necessary. For instance, in relation to the home model (see Figure 2.7), the object-related processes of a dishwasher may be represented initially as *washing capacity*, and specific

---

* See Chapter 2 for discussions that justify the recommended checks for constraints.

BOX 8.7D   HINTS FOR CONSTRAINTS

- Four questions are worth considering when specifying the constraints to include in the model of the focus system: What is the potential content of the cells in the model? Which parts of the system at each level of decomposition should be represented with constraints? Which cells of the model should be populated with constraints? and What constraints should be portrayed in the selected cells? These questions are not necessarily dealt with in a discrete, linear order, especially by experienced analysts.

- A skeletal abstraction–decomposition space of the focus system revealing the potential content of the cells in the model is useful for deciding which parts of the system to represent with constraints, which cells of the model to populate with constraints, and which constraints to include in the selected cells.

- Analysts planning to model the focus system with an abstraction hierarchy should consider assembling a skeletal abstraction–decomposition space by analyzing a few parts of the system, at least, to formulate the decomposition dimension of the matrix.

- In deciding which parts of the system to model with constraints, two options are the first and second variations of system decomposition (see Figures 4.6a and 4.6b).

- In selecting which cells of the model to populate with constraints, it is sensible to start by evaluating the cells around the diagonal. Having established which of these cells offer the most significant information for the purpose of the analysis, the remaining cells may be examined to determine whether they offer additional information that is relevant for the proposed application.

- The constraints that are included in the model of the focus system are determined by the levels of abstraction and decomposition of the selected cells and the nominated parts at each level of decomposition.

- If the levels of abstraction in the model are the same as those in the standard set, the prompts and keywords in Figure 8.7 and the detailed descriptions of the standard levels in Chapter 3 may be used for defining constraints. In addition, irrespective of whether the levels of abstraction are the same as or different from those in the standard set, the prompts and keywords in Table 8.1 and the structural properties of the environment uncovered during the process of specifying the abstraction dimension without presuming the standard levels may be beneficial for identifying constraints.

- After one or more constraints have been revealed for the model of the focus system, structural means–ends relations also provide a basis for specifying constraints.

- The model may be supplemented with a glossary that describes the constraints in more detail.

- In defining the constraints to portray in a model, it is worth checking that (1) the constraints within a level of abstraction signify the same concepts in the same modeling language, whereas those at different levels reflect different concepts in distinctive languages; (2) the relationships between the constraints are structural means–ends relations; (3) the constraints represent the work domain at the same or different resolutions, depending on whether they occur at the same or different levels of decomposition; and (4) the constraints at higher levels of decomposition provide coarser representations of the work domain than those at lower levels.

- When specifying the constraints to represent at the standard levels of abstraction, it may be easiest to begin at the topmost and bottom two levels.
- Within a level of abstraction in the model of the focus system, constraints may be represented at one of several levels of decomposition. Analysts need to be careful to place constraints at the appropriate level of decomposition.
- The constraints in a model should be specified to a level of depth or completeness that is suitable for the proposed application. If there is some uncertainty about the requirements of the application, it may be sensible to begin by developing the model broadly and then add information as necessary.
- Analysts need to be careful that they do not represent the same constraints multiple times in a model, perhaps in different guises or as various instances of system functioning.
- The original manifestations of the structural properties of the environment, as revealed by the information sources, may need to be recast in a way that emphasizes the constraints on actors' behavior.
- *How* and *why* questions may be employed to establish whether the constraints in the model of the focus system are defined by structural means–ends relations. These questions, however, are likely to invoke instantiations of means–ends relations, rather than the full set of possible relationships (see Chapter 5), and slight variations in the questions may give rise to different instantiations.
- Structural means–ends relations and part–whole relations are fundamental to the abstraction–decomposition space. However, sometimes it may be useful to represent topological relations in the model as well, depending on the purpose of the analysis (see Chapter 5).
- Vicente (1999) emphasizes that work domain analysis is a highly iterative process, and that analysts must be prepared to revise their models many times.

details about its capacity (e.g., load limits, washing cycles, temperature range) may be added as required.

The information sources for an analysis may reveal the same constraints in many different guises or in the form of a range of instances of system functioning.[*] For example, with respect to the model of a home, properties such as money, income, expenses, and savings may all signify the same constraint, that is, *total income – total expenses = savings*. Similarly, numerous instances of system functioning referring to various sorts of meals prepared in the home may all denote the same constraint, specifically *preparation of meals and beverages*. Analysts must be careful not to represent the same constraints multiple times in a model in a variety of manifestations.

The preceding point also highlights that the information sources may not reveal the structural properties of the environment in a form that is suitable for representing constraints. Therefore, the original expressions of these properties may need to be modified for the model of the focus system. For example, in the case of a home, a property revealed

---

[*] See Theme 6 in this chapter for a description of instances of system functioning.

as 'prepare meals and beverages' (Figure 8.8a) may be better portrayed as *preparation of meals and beverages* so that it more clearly signifies a function that the home must be capable of supporting to achieve its functional purposes, instead of actors' behavior. In the same way, properties such as 'cooling' and 'heating' may be better described as *cooling capacity* and *heating capacity* as these labels more clearly denote functional capabilities of physical objects rather than actors' behavior.

How and *why* questions may be helpful for verifying that the constraints in the model of the focus system are characterized by structural means–ends relations (Vicente, 1999). That is, of each constraint in the model, analysts may ask whether constraints at lower levels of abstraction explain *how* that constraint can be attained and, conversely, whether constraints at higher levels explain *why* that constraint must be achieved. Bear in mind, though, that this process is most likely to invoke instantiations of means–ends relations and not the full set of means–ends relations that is possible.[*] Furthermore, posing different kinds of how and why questions, such as asking why a constraint must be achieved, why a constraint is relevant, or why a constraint is necessary, may give rise to different instantiations.

Theme 8: Is the Abstraction–Decomposition Space a Valid Model of the Focus System?

Also critical to address when performing a work domain analysis is the question of whether the abstraction–decomposition space is a valid model of the focus system. In other words, given the purpose of an analysis, does the abstraction–decomposition space provide a comprehensive and accurate representation of the focus system? This is an important matter to settle because it will lead either to recognition of the need for improvements to the model or to confidence in the model so that the development of designs or solutions for workers can begin.

Typically, analysts have employed two strategies for validating a work domain model. The first involves reviewing the model with subject matter experts. The second entails mapping scenarios onto the model.

Prior to presenting these strategies, it is important to acknowledge that it is difficult to validate a work domain model fully or completely. Small amounts of missing information or minor inaccuracies in the model may be discovered with every validation exercise. Even if this were not the case, one could never be sure whether reviewing the model with yet another subject matter expert, or testing the model with yet another scenario, would uncover further problems. Nevertheless, at some point the validation process will need to be terminated, not least because of the project restrictions. A judgment is required, therefore, as to when it is sensible to proceed with developing solutions from the model.

Bear in mind also that, ultimately, a model must be evaluated with respect to its usefulness (Burns and Hajdukiewicz, 2004; Vicente, 1999), that is, whether it is useful for the purpose for which it was developed or perhaps for other applications. In some cases, it may be possible to establish the usefulness of the model by evaluating the performance

---

[*] Chapter 5 explains the difference between instantiations of structural means–ends relations and the full set of possible means–ends relations.

of workers using a solution generated from the model. For example, laboratory or field experiments may be conducted to determine whether users of an ecological interface perform better than users of a conventional interface. In other cases, the usefulness of the model may be established by its impact, or ability to influence practice, and its uniqueness (Naikar, 2009). For instance, one may evaluate whether the resulting solution is adopted by workers and whether it is unique in comparison with those produced by other approaches. Chapters 9, 10, and 11 demonstrate how the usefulness of a model may be evaluated in these ways.

Review with Subject Matter Experts

One strategy for checking the validity of a work domain model is to have it reviewed by subject matter experts. This strategy does not necessarily involve showing experts the model but, instead, may involve asking them questions about aspects of the model, including its levels of abstraction and decomposition, structural means–ends and part–whole relations, and constraints. Experts may confirm the content of the model or highlight deficiencies in the model, such as missing or inaccurate information.

Consider, for instance, the model of a home (Figure 2.7). Experts may be presented with a list of the rooms or subspaces in the model and asked to confirm whether those items are in fact rooms or areas of the house and whether they can identify any others. Likewise, they may be shown the purpose-related functions and requested to verify whether those functions capture how the various areas of the home can be used and whether they can think of any other actual or potential uses. Similarly, experts may be provided with the set of value and priority measures and asked to indicate which items are considerations that influence the preparation of meals and beverages in the home and whether there are any other factors. Clearly, many other questions are possible. The extensiveness of the review, in these and other respects, will depend largely on the project restrictions.

It is useful to involve a variety of experts in reviewing the model, as is the case for the process of building the model. In Theme 6, it was noted that a wide range of experts may be consulted for work domain analysis, including directors, executives, strategists, legislators, administrators, managers, frontline workers, training instructors, maintenance personnel, scientists, engineers, and system developers. Together, different experts may possess the breadth of knowledge or experience required to establish the validity of the model.

The aspects of the model that are given to experts for review may depend on the nature of their expertise. As a case in point, engineers or maintenance personnel may be shown details about the system's tools or equipment, represented at lower levels of abstraction in the model, whereas directors or executives may be asked to consider information about the system's objectives or values, depicted at higher levels of abstraction. The language or wording of the questions put to experts may also be tailored to their expertise.

Questions about the model may be posed to experts in different ways. For example, experts may be asked questions in the form of interviews, tabletop analyses, or questionnaires. As tabletop analysis allows experts to participate in the review as a group, this technique has the advantage of bringing together people with different perspectives to assess the model.[*]

Different options are also available for presenting the model, or aspects of it, to experts for examination. For instance, experts may be provided with information about the model as a written list or verbal description. In some cases, showing experts the entire model, or

---

[*] More information about tabletop analysis is provided in Theme 6 of this chapter.

elements of it, in a graphical or tabular format may be appropriate. Scientists, engineers, or system developers, for example, are usually comfortable with working with these kinds of representations.

Burns et al. (2005) report a validation exercise in which experts were asked to review a model of a naval command-and-control system, specifically a frigate.* The experts were operations room officers, who were part of the command team of the frigate. The officers were given a questionnaire in which they were asked to "review a list of work-domain elements and verify whether or not these elements were part of the work domain" (Burns et al., 2005, p. 614; also see Burns and Hajdukiewicz, 2004). Although the officers confirmed the appropriateness of various aspects of the model, Burns et al. found that the review was not as useful for highlighting information that was missing from the model as was a scenario-mapping exercise, which is discussed in the next section.

Scenario Mapping

Scenario mapping is another strategy for establishing the validity of a work domain model (Burns, Bryant, and Chalmers, 2001, 2005; Burns and Hajdukiewicz, 2004; Miller and Vicente, 1998). Mapping a scenario onto the model of the focus system involves examining whether the model can be used to understand, explain, or predict the system's responses in that scenario. This strategy recognizes that a work domain model is event independent.† Accordingly, it should be possible to use such a model to reason about the behavior of the system in any situation, including novel or unanticipated events. By assessing the capacity of the model to reason about the system's responses in various situations, a scenario-mapping exercise can verify the content of the model or reveal problems with the model, such as missing or inaccurate information.

Scenario mapping may be implemented in a variety of ways. The rest of this section describes two studies that demonstrate some of the possible variations. One study used subject matter experts to perform the scenario mapping and adopted a familiar scenario. The other study used analysts to conduct the scenario mapping and utilized a novel scenario.

As well as the review with experts described previously, Burns et al. (2001, 2005) held a scenario-mapping exercise to validate the model of the frigate. Three operations room officers, belonging to the frigate's command team, performed the scenario mapping. An existing training scenario, which was familiar to the officers, was used for the exercise. As this scenario had not been employed for model development, it was suitable for use in the validation process.

During the exercise, the experts were given copies of the model and an outline of the events in the scenario. Working individually, the experts examined each event and, on the basis of the actions that would be required of them when dealing with each event, they indicated which of the constraints in the model "were in use at each step, and any other information that they use" (Burns et al., 2001, p. 425). Following the consideration of each event, the analysts held a joint discussion with all of the experts "to compare opinions and extract more information with respect to the scenario" (Burns et al., 2001, p. 425). At the end of the session, the analysts evaluated the full set of information provided by the experts against that contained in the model.

---

* See Figures 6.9, 6.10, and 6.11 for different aspects of the model of the frigate.
† The concept of event independence is described in Chapter 2.

BOX 8.8   IS THE ABSTRACTION–DECOMPOSITION SPACE A VALID MODEL OF THE FOCUS SYSTEM?

- Given the purpose of the analysis, does the abstraction–decomposition space provide a comprehensive and accurate representation of the focus system?

HINTS

- One strategy for validating a work domain model is to have it reviewed by subject matter experts. Another is scenario mapping.
- The extent or scope of the validation exercise, in the case of either strategy, may depend on the project restrictions (see Theme 2).

According to Burns et al. (2001, 2005), the scenario-mapping exercise confirmed that the model was reasonably complete. Information that was found to be missing from the model was added to it. Although some of the constraints in the model were not relevant during the scenario, this finding did not necessarily mean that these constraints were erroneous and should be removed from the model. Instead, these constraints might be important in other scenarios. Burns et al. observed that this type of exercise, therefore, was limited in its capacity to reveal inaccuracies in the model.

Miller and Vicente (1998) also performed a scenario-mapping exercise to validate a model of an acetylene hydrogenation reactor in a petrochemical plant. In this case, the scenario was a real incident. Moreover, the scenario was novel at the time of its occurrence in the sense that it was unfamiliar to operators and had not been anticipated by experts. Although the incident had occurred some time prior to the development of the model, it was not referred to in the plant's normal operating procedures or any other documentation that had been used for model construction.

To validate the model, Miller and Vicente (1998) studied the sequence of events in the incident, as described in an incident investigation report, and examined whether the model could be used to explain the causes of the incident, the structural states produced in the plant during that time, the motivations of operators as they attempted to address the situation, and the effects of operators' actions on the plant. They found that the model could explain the incident satisfactorily, although aspects of the model were either incomplete or inaccurate. In addition, the model could be used to make valid predictions about the effects of various problems during the incident and to make recommendations for the management of those problems. Hence, it was plausible that operators would have found a display based on the model helpful for diagnosing and responding to the situation.

Summary

In this chapter, guidelines for performing work domain analysis were presented in the form of a set of analytic themes or questions (Box 8.9), which emphasizes the flexible and iterative nature of this approach. These themes were conceived on the basis of three key considerations, specifically the underlying concepts of work domain analysis, numerous

BOX 8.9   ANALYTIC THEMES FOR WORK DOMAIN ANALYSIS

THEME 1: WHAT IS THE PURPOSE OF THE ANALYSIS? (SEE BOX 8.1)

- What objective do I want to achieve with work domain analysis?
- How will I use a work domain model to fulfill that objective?

THEME 2: WHAT ARE THE PROJECT RESTRICTIONS? (SEE BOX 8.2)

- What is the nature of the project restrictions?
- How should I perform the work domain analysis given those restrictions?

THEME 3: WHAT ARE THE BOUNDARIES OF THE ANALYSIS? (SEE BOX 8.3)

- What is the focus system for the analysis? That is, which organizational entity, physical entity, problem, or actors' perspective should the analysis focus on?

THEME 4: IS IT USEFUL TO DEVELOP MULTIPLE MODELS? (SEE BOX 8.4)

- Is it useful to develop multiple models to capture different stakeholders' perspectives of a problem?
- Is it useful to develop multiple models to emphasize the distinct facets of a problem or the fundamental differences in control that a system has over elements of its work domain?

THEME 5: WHERE ON THE CAUSAL–INTENTIONAL CONTINUUM DOES THE FOCUS SYSTEM FALL? (SEE BOX 8.5)

- What is the focus system's position on the causal–intentional continuum? That is, is actors' behavior shaped primarily by intentional or causal constraints?
- What is the nature of the constraints that should be represented in a model of that system?

THEME 6: WHAT ARE THE SOURCES OF INFORMATION FOR THE ANALYSIS? (SEE BOX 8.6)

- What documents, field observations, or subject matter experts are useful for gathering information about the focus system?

THEME 7: WHAT IS THE CONTENT OF THE ABSTRACTION–DECOMPOSITION SPACE? (SEE BOX 8.7)

- What levels of abstraction should I include in the model of the focus system?
- What levels of decomposition should I incorporate into the model?
- What constraints should I represent in the model?

THEME 8: IS THE ABSTRACTION–DECOMPOSITION SPACE A VALID MODEL OF THE FOCUS SYSTEM? (SEE BOX 8.8)

- Given the purpose of the analysis, does the abstraction–decomposition space provide a comprehensive and accurate representation of the focus system?

case studies of the application of this form of analysis, and evaluations of earlier versions of the guidelines. Each theme was illustrated with examples from a home.

In the following chapters, Chapters 9, 10, and 11, the guidelines are illustrated by three instances of their application to large-scale military systems in industrial settings. These case studies serve the purpose of providing more complex demonstrations of the guidelines. In addition, these case studies emphasize that work domain analysis has a range of applications in industrial settings, beyond interface design. Specifically, it can be used to evaluate design concepts (Chapter 9), develop team designs (Chapter 10), and examine training needs and instructional system requirements (Chapter 11). For the first time, the usefulness and feasibility of work domain analysis for these industrial applications are considered in detail.

Section IV

Cases

9

Evaluation of Design Concepts

OVERVIEW   Work domain analysis provides a powerful approach for addressing a variety of industrial problems aside from designing ecological interfaces. This chapter describes its use to evaluate design concepts for complex sociotechnical systems. After explaining the significance of this problem, and the suitability of this form of analysis for addressing it given the characteristics of complex sociotechnical systems, a comprehensive case study is presented. This case study shows how work domain analysis was implemented to assist the Australian Defence Organisation in evaluating three design concepts for the Airborne Early Warning and Control system. Specifically, it describes how the analysis was performed, in accordance with the analytic themes in Chapter 8, and how the resulting model was used to assess the design concepts. Following that, the usefulness and feasibility of work domain analysis for this application are examined, and some limitations of this case study are discussed.

Context

The procurement of complex military systems, such as fighter jets, armored land vehicles, or warships, is a lengthy and challenging undertaking. The point of this chapter is to investigate the applicability of work domain analysis to one of the stages of the military acquisition cycle, specifically the evaluation of alternative design concepts for the proposed system. This stage is called 'tender evaluation' in the United Kingdom and Australia and 'source selection' in the United States of America. The next two chapters are concerned with the appropriateness of work domain analysis for attending to other phases of the procurement process: team design and training needs analysis, respectively. Although these applications of work domain analysis are presented in the context of military systems, the technique can be used to evaluate design concepts, develop team designs, and examine training needs and instructional system requirements for any type of system, not just military systems.

   Ultimately, the task of assessing alternative design concepts for the proposed system has as its objective the selection of the best design. In the case of military procurement, the candidate designs are usually put forward by competing industry manufacturers. Once a contract has been established with the winning manufacturer, the procurer becomes committed to a specific design concept. Further evaluations of the system design, later in the acquisition cycle, are mainly concerned with refining the initial concept. This means that reaching a sound decision about the best design for the intended system during the tender evaluation or source selection stage is crucial.

Rationale

The characteristics of complex sociotechnical systems create special requirements for the evaluation of designs. These systems comprise an interdependent set of physical components or devices, which are necessary for achieving their work demands. This means that a framework for evaluation must be concerned with assessing not only individual physical devices but also multiple, interacting physical devices. Furthermore, the criteria for evaluation must be aimed at establishing whether the designs of the set of physical devices will fulfill the system's work demands.

The work demands of complex sociotechnical systems cannot be specified solely in the form of stable sets of task sequences or procedures (Meister, 1996; Rasmussen et al., 1994; Vicente, 1999). Such descriptions can only be conceived in relation to events that are known or can be anticipated.[*] The work demands of unforeseen events cannot be stated in these terms because the goals in these situations are not known a priori. Novel events, however, pose the greatest threats to a system's performance and safety (Perrow, 1984; Rasmussen, 1969; Reason, 1990; Vicente, 1999). Typically, workers cannot rely on standard work practices for handling these situations. Instead, flexible or innovative behavior is usually necessary. A framework for evaluation, therefore, must accommodate judgments of how well a proposed design will support workers in managing the demands of a range of situations, including those that are unfamiliar or unanticipated.

Work domain analysis provides an approach for evaluating design concepts for complex sociotechnical systems that is well suited to their special characteristics (Naikar and Sanderson, 2001). To explain this proposition, Figure 9.1 presents a generic abstraction hierarchy comprising the standard five levels of abstraction (Rasmussen, 1986; Rasmussen et al., 1994).[†] Using this approach, a system's work demands may be expressed in terms of the functional purposes, value and priority measures, and purpose-related functions that must be satisfied with a suite of physical resources. It then becomes possible to assess design concepts with respect to how well these work demands are supported. Specifically, the technical solutions of a design concept, which encompass the proposed physical devices and their functional capabilities and limitations, can be mapped onto the lowest two levels of abstraction, as shown in the figure. Following that, the technical solutions can be assessed in terms of their adequacy for fulfilling the purpose-related functions, value and priority measures, and functional purposes.

For example, as highlighted by the shaded nodes in Figure 9.1, Physical Device B and Functional Capability and Limitation B can be examined in terms of their effects on Purpose-related Functions Y and Z. The impact of this technical solution on the purpose-related functions can then be considered with respect to its implications for Value and Priority Measures V and W and the Functional Purpose of the system. It is also possible to take into account the impact of the interactions between the technical solutions on the system's work demands. As a case in point, the interactions between Physical Devices B and C, and Functional Capabilities and Limitations B and C, can be assessed in terms of their consequences for Purpose-related Function Z and Value and Priority Measure W. One possibility is that both technical solutions have negative effects on Purpose-related

---

[*] As discussed in Chapters 2 and 7, action means–ends relations can be specified only for known or anticipated events.

[†] Detailed descriptions of the standard five levels of abstraction are provided in Chapter 3.

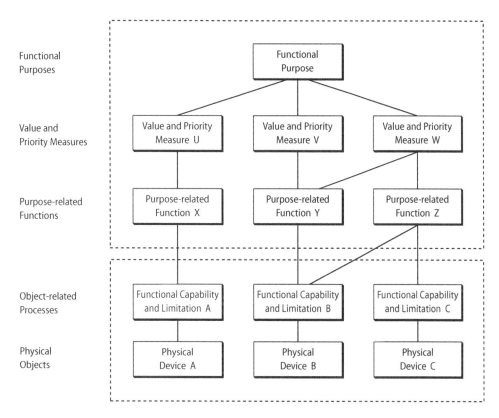

FIGURE 9.1
Generic illustration of the application of work domain analysis for evaluating design concepts.

Function Z, which compromise Value and Priority Measure W. Another possibility is that the two technical solutions have positive effects on Purpose-related Function Z, which enhance Value and Priority Measure W. Alternatively, one of the technical solutions may have a positive effect on Purpose-related Function Z while the other has a negative effect on this function, such that the effect of one is canceled by the other and, overall, there is no impact on Value and Priority Measure W.

This framework for evaluation is well suited to complex sociotechnical systems because designs can be assessed with respect to whether an interdependent set of technical solutions would fulfill a system's work demands. Furthermore, the work demands are characterized in the form of a set of constraints, which are event independent.* Specifically, these constraints stipulate the functional purposes, value and priority measures, and purpose-related functions that must be attained with a suite of physical resources, irrespective of the situation. Therefore, by assessing a design concept against these constraints, it becomes possible to develop an understanding of how well it will support workers in dealing with the demands of an assortment of situations, including unexpected events.

---

* Chapter 2 discusses the event independence of a work domain model in more detail.

Case Study

This case study describes how work domain analysis was applied to evaluate three design concepts for the Airborne Early Warning and Control (AEW&C) system, valued at approximately Aus$3 billion, for the Australian Defence Organisation. The three design concepts were submitted by competing industry manufacturers. The AEW&C system was expected to comprise a set of six aircraft, each operated by a pilot and a copilot, referred to as the flight crew, and a team of people in the cabin of the aircraft, called the mission crew. Its main roles would be to develop a tactical picture and coordinate the activities of various defense assets, such as fighter aircraft and maritime patrol aircraft, in an allocated area of operations. To perform these roles, the aircraft would be fitted with a suite of physical devices, including sensors, computers, voice and data communication systems, and electronic warfare equipment. This system may be likened to the Airborne Warning and Control System (AWACS) of the U.S. Air Force or the E2C system of the U.S. Navy.

Work Domain Analysis

The work domain analysis of the AEW&C system, from now on mostly referred to as AEW&C, may be described according to the analytic themes in Chapter 8. These themes had not been formulated explicitly when this analysis was conducted. However, the analysis can be easily conceptualized in these terms as it did, in fact, inform the development of the themes. Readers who are not interested in the details of how this analysis was performed may proceed to the next section, which discusses how the resulting model was used to evaluate the design concepts.

Theme 1: Purpose of the Analysis

The objective of the analysis was to help the Australian Defence Organisation evaluate the three design concepts for AEW&C and, subsequently, select the best design concept for this system, specifically for the aircraft. The plan for how the model would be used to fulfill this objective was described in general terms previously.

Theme 2: Project Restrictions

The project restrictions related to schedule, personnel, and finances. In what follows, the nature of the main restrictions is described. The impact of these restrictions on how the analysis was carried out becomes evident in the discussion of the subsequent themes.

Approximately one year was available for performing the analysis and using the resulting model to evaluate the design concepts. The analysts available for developing the model were myself (a researcher at the Defence Science and Technology Organisation) and a university-based consultant. About half of my time could be allocated to this project over a one-year period. Although I had not conducted a work domain analysis previously, the consultant had experience with this type of analysis, but not on such a complex system and not in an industrial setting.

A recently retired military operator was contracted as a dedicated subject matter expert. This person had four years of operational experience in an Australian ground-based system with similar roles to AEW&C. In addition, she had considerable knowledge of the proposed operational and system concept for AEW&C, having worked for three years in the AEW&C System Program Office, the main defense unit responsible for the acquisition of this system.

The personnel available for using the resulting model to evaluate the design concepts included approximately one hundred scientists, engineers, and military specialists, predominantly from the Australian Defence Organisation. These personnel had the requisite technical and military expertise for establishing how well the technical solutions of the design concepts satisfied the work demands of AEW&C.

Theme 3: Boundaries of the Analysis

The organizational entity that framed the boundaries of the analysis was the Australian Defence Organisation's future AEW&C system. The physical entity was a single aircraft. Although AEW&C comprises a set of six aircraft, the analysis could be restricted to a single aircraft because the work demands of the system are the same for every aircraft. In addition, the boundaries were drawn around the problem of the development of a tactical picture and the coordination of defense assets in an allocated area from the perspective of actors operating the aircraft.

It is important to note that AEW&C consists of components other than the aircraft, such as ground-based mission planning and training facilities. While it may have been beneficial to incorporate all of these components into the model, and thus into the evaluation of the design concepts, this was not possible because of the project restrictions. Instead, the analysis focused on the aircraft because the design of this component posed the most significant risk to the effectiveness of AEW&C. This does not mean that the other components were not subjected to any kind of evaluation—they were assessed using other methods, which are described briefly later in this chapter.

Trade-offs were also necessary with respect to the problem that was selected as the focus of the analysis. In particular, the analysis was directed at the problem of the development of a tactical picture and the coordination of defense assets in an allocated area rather than that of flying the aircraft. This decision reflects the fact that key technical solutions for supporting the focus problem, including sensors and automated situation and threat assessment devices, were considered to be of high risk to the effectiveness of AEW&C. These technical solutions were revolutionary in their design compared with those for current-generation systems, namely, AWACS and E2C. In contrast, the technical solutions central to the problem of flying the aircraft, such as engines and navigation devices, were considered to be of relatively low risk as they were similar to those in a range of military and commercial aircraft, including AWACS and E2C. These technical solutions were still represented in the model, since they are part of the aircraft, but they were evaluated predominantly in terms of how well they supported the work demands associated with the focus problem.

Theme 4: Multiple Models

Multiple models of stakeholders' perspectives could have been created for AEW&C. Specifically, the focus problem could have been portrayed separately from the perspectives of the flight crew and the mission crew. The flight crew's stake in this problem concerns such activities as navigating or positioning the aircraft appropriately, whereas the mission crew's interest in this problem involves such activities as locating or identifying other entities in the area. By providing detailed representations of the work demands of the two crews in relation to this problem, multiple models could have led to a detailed understanding of how the technical solutions of the design concepts affect the work demands of each crew. However, as this approach was not feasible given the project restrictions, a single model that encompassed the perspectives of both crews, conceptualized as actors operating the aircraft, was assembled. Such a model still promoted an understanding of the effects of the technical solutions on both crews, albeit at a less detailed level.

Multiple models could also have been formulated to emphasize the distinct facets of the focus problem since AEW&C is a loosely bound system. While the system has control over the resources for managing threats (e.g., friendly fighter aircraft), it does not have control over the potential sources of threat (e.g., enemy fighter aircraft) or the potential targets of damage (e.g., civilian population and infrastructure). In addition, it has responsibility for handling disturbances that arise externally (e.g., enemy attacks). Therefore, by differentiating risks from resources for managing those risks, two models could have been developed. The model of the domain of potential risk would have included the potential sources of threat and the potential targets of damage, whereas the model of the domain of risk management resources would have included the resources for managing threats. Such a representation would have promoted a thorough understanding of how well the technical solutions of the design concepts support risk assessment and risk management.

However, multiple models of problem facets were not constructed for AEW&C. A single model that showed how the physical objects and object-related processes of the system contributed to its purpose-related functions, value and priority measures, and functional purposes was better suited to the purpose of the analysis. This model could be used to trace the impact of the technical solutions for the aircraft, represented at the lower two levels of abstraction, on the constraints at the higher levels. As the potential sources of threat, potential targets of damage, and resources for managing threats were integrated implicitly into the representation of the constraints at each level, this model still portrayed that the focus problem involves risk assessment and risk management. The single model, therefore, did lead to an understanding of how well the technical solutions support risk assessment and risk management, although the evaluation may not have been as detailed and explicit as would have been the case with multiple models. On the other hand, as the interactions between threats, targets, and resources were more apparent in the single model, it probably led to a better appreciation of the impact of the technical solutions on the work demands associated with these interactions. Admittedly, it may have been beneficial to develop multiple models in addition to the single one, but this was not possible given the project restrictions.

Theme 5: Causal–Intentional Continuum

The focus system was found to fall toward the right of the middle of the causal–intentional continuum, so that it was classified as a loosely coupled intentional system.[*] This means that actors' behavior is shaped to a greater degree by intentional constraints relative to causal constraints. Intentional constraints were evident in organizational policies (e.g., Air Force doctrine), plans (e.g., mission plans), legislation (e.g., international military law), and regulations (e.g., rules of engagement). Causal constraints had their basis in loosely coupled physical devices present on the aircraft and the functional capabilities and limitations of those devices. Within the bounds of the mission objectives, actors in this system have reasonable flexibility in setting local goals and priorities as well as considerable control over how physical resources are used to achieve the high-level objectives. Consequently,

---

[*] See Figure 8.5 for a graphical depiction of the causal–intentional continuum.

the model of the focus system was characterized predominantly by intentional constraints at higher levels of abstraction and causal constraints at lower levels.

Theme 6: Sources of Information

As AEW&C is a future system, the main information sources for the analysis were documents and subject matter experts. The documents included a concept of operations, summary of requirements, system specification, use study, and a range of technical reports covering the physical components of the system. The *Air Power Manual*, which describes the Royal Australian Air Force's doctrine, was also reviewed for the analysis.

All of the subject matter experts had considerable knowledge of the proposed operational and system concept for AEW&C. They included military personnel with operational experience in current-generation systems with comparable roles to AEW&C, specifically AWACS and E2C; military personnel with operational experience in an Australian ground-based system with similar roles to AEW&C; senior defense personnel from the Australian Defence Organisation with broader, strategic knowledge relating to AEW&C; and a variety of engineers, operations analysts, and scientists from the Defence Science and Technology Organisation with extensive technical knowledge of AEW&C.

The knowledge-elicitation sessions with experts mainly took the form of talkthroughs, unstructured and semistructured interviews, and table-top analyses as these techniques are suitable for gathering information about future systems. We also attended a two-day tutorial on the Australian ground-based system, which was delivered by military personnel with relevant operational experience. This tutorial was useful primarily for learning about the class of system to which AEW&C belongs.

Some field observations were also conducted for the analysis, but, like the tutorial, they were limited in the information they could offer about AEW&C. These included a site visit to the Australian ground-based system, where it was possible to observe simulated operational exercises. In addition, an opportunity arose to conduct observations onboard an AWACS aircraft during an international military exercise held in Australia. Numerous technical meetings concerning the various physical components and operations of AEW&C provided more specific information about this future system.

Theme 7: Content of the Abstraction–Decomposition Space

The levels of abstraction and decomposition and the constraints in the model of AEW&C were specified as follows. The standard set of five levels of abstraction (Rasmussen, 1986; Rasmussen et al., 1994) was adopted for the model.* The labels given to these levels changed throughout the project but were mostly variations of those used by Rasmussen and his colleagues. For the sake of simplicity, I refer to these levels with the labels most recently suggested by Rasmussen, which are functional purposes, value and priority measures, purpose-related functions, object-related processes, and physical objects (Reising, 2000).

The decomposition dimension of the model was defined using constraint decomposition.† That is, once a set of constraints had been identified, the constraints were decomposed or aggregated into respectively more or less detailed representations of the work

---

* I would now lean toward defining the abstraction dimension of a model without assuming the standard levels. The process for doing so is described in Chapter 8.
† I would now always adopt system decomposition because it enables a skeletal abstraction–decomposition space of the focus system to be constructed, which has many benefits, as outlined in Chapter 4.

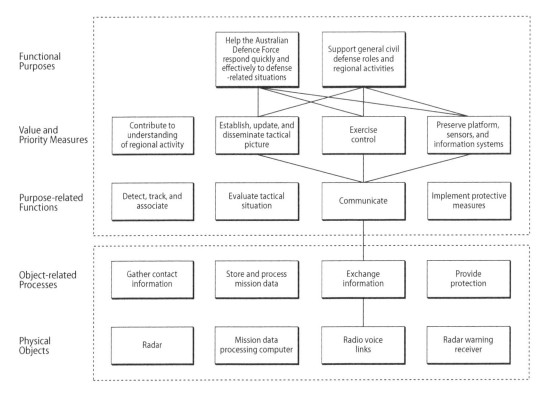

FIGURE 9.2
Sample of constraints and structural means–ends relations in the AEW&C model.

domain. For example, one of the purpose-related functions of AEW&C was *evaluate tactical situation*. This constraint was decomposed into types of tactical information, including information about the tracks or entities in the environment, the weather, and the terrain.

As system decomposition had not been adopted for the model, the constraints were not specified using the full process described in Chapter 8. Instead, knowledge of the concepts that characterize the standard levels and structural means–ends relations were used to define the constraints.

Information about the constraints pertinent to AEW&C was obtained from documents, field observations, and subject matter experts. Generally, details about the functional purposes of the system were revealed in explanations about why it was needed, or why it was being acquired by the Australian Defence Organisation, and how it would contribute to Australia's defense capability. The value and priority measures were derived from explanations about the advantage or edge AEW&C was expected to offer over potential adversaries. The purpose-related functions were disclosed in explanations about the typical functions its crew would perform on missions. Last, the object-related processes and physical objects were evident in explanations about the physical devices that were necessary onboard the aircraft given the intended roles of the system.

Figure 9.2 presents a sample of the constraints in the model.* In the interests of legibility and interpretability, the labels of some of the constraints have been edited slightly. The functional purposes level shows that AEW&C will contribute to military defense

---

* I would now represent constraints with labels that make the functional structure of the environment of actors more apparent.

operations by helping the Australian Defence Force respond quickly and effectively to threats. It will also assist with civil defense roles, such as maritime surveillance, and play a part in regional activities, such as disaster relief. To achieve these functional purposes, the system must satisfy a number of value and priority measures, including a long-term understanding of patterns of regional activity, a near real-time understanding of the current tactical situation, the safety and effectiveness of defense assets under its control, and self-preservation. The purpose-related functions encompass the development of a tactical picture, evaluation of the tactical situation, communication, and implementation of protective measures. At the lowest two levels, the physical objects and object-related processes include a radar for gathering information about entities in the environment, a mission data processing computer for storing and processing information about the mission, radio voice links for exchanging voice information, and a radar warning receiver for providing protection by presenting warnings of incoming radars.

Figure 9.2 also shows some of the structural means–ends relations in the model. These links may be used to provide a simple example of how the model can be used to consider the impact of the technical solutions of a design concept on the work demands of AEW&C. For instance, if there are deficiencies in the design of the *radio voice links* such that the capacity to *exchange information* and *communicate* is limited, then the ability to *establish, update, and disseminate tactical picture* and *exercise control* over friendly assets will be compromised. These effects, in turn, will propagate to both of the functional purposes. However, if the shortcomings in the design of the *radio voice links* are such that radio emissions from the aircraft are reduced (*exchange information*), then the presence of the aircraft will be relayed less broadly (*communicate*), which will help to *preserve platform, sensors, and information systems* from attack.

It is important to emphasize that the preceding example is a hypothetical illustration that has been deliberately kept simple to suit readers with little knowledge of military systems. It was not an output of the evaluation of the design concepts for AEW&C. More complex examples, which are also necessarily hypothetical for commercial and security reasons, are presented in the next section. In particular, these examples demonstrate how the model can be used to judge the impact of multiple, interacting technical solutions on the work demands of AEW&C.

Finally, a glossary was developed to supplement the model, which was depicted in the form of both a graphical abstraction hierarchy and a tabular abstraction–decomposition space.* The glossary described the constraints in the model in greater depth than did the graphical or tabular presentations. It also explained the reasoning behind the structural means–ends relations between particular constraints.

Theme 8: Validity of the Model

The strategy for validating the model was to have it reviewed by subject matter experts. These experts had similar proficiencies to those who were consulted in the development of the model. The review involved asking experts various questions about the content of the model during semistructured interviews or tabletop analyses. The features of the model they were asked to review were tailored to their knowledge and experience. Some of the experts were shown the model during the review. For others, aspects of the model were presented in the form of verbal descriptions.

---

* Chapter 2 discusses formats for presenting work domain models.

The review verified many elements of the model and also highlighted some minor problems with it, such as inaccurate or missing information, which led to the model being refined. The final model comprised 2 functional purposes, 10 value and priority measures, 12 purpose-related functions, 77 object-related processes, and 61 physical objects. Approximately 350 structural means–ends relations were established between these constraints, covering such instantiations as those that would be engineered into the system, those that were known or could be anticipated by experts, and those that were identified in the context of specific situations.[*]

Evaluation of Design Concepts

This section describes how the AEW&C model was used to evaluate the three design concepts for this system, specifically for the aircraft. Before proceeding, it is important to point out that two other approaches for evaluation were also implemented on this project, that is, a technical evaluation and an operational evaluation. The technical evaluation involved assessing whether the technical solutions of the design concepts complied with prespecified functional requirements, surpassed the requirements, or were deficient in some regard. An example of a functional requirement for AEW&C is that the aircraft's radar must be able to detect entities in the environment at a range of 300 nautical miles. For the operational evaluation, computer-based models of aspects of the technical solutions of the design concepts were developed, and then the performance of these models was tested in a set of six scenarios using Monte Carlo simulation. A general description of a scenario for AEW&C is 'conduct and control a maritime strike over blue-water ocean.'

The process by which the AEW&C model was used to assess the design concepts took advantage of the structure of the team for the technical evaluation. As shown in Figure 9.3, this team was organized according to subgroups that were responsible for reviewing certain technical solutions for the aircraft. For example, the radar subgroup was concerned with technical solutions for the radar; the electronic warfare subgroup was interested in technical solutions for the electronic warfare devices, which included a radar warning receiver; and the mission system subgroup was focused on technical solutions for the mission system, which included a mission data processing computer. Each subgroup consisted of scientists, engineers, and military personnel with suitable technical and military expertise for appraising the technical solutions for which they were accountable. Every subgroup had a leader who reported to the head of the evaluation team.

Initially, the subgroups conducted the technical evaluation. That is, the technical solutions of the design concepts were assessed against the prespecified functional requirements. Subsequently, the subgroups and a panel, which comprised the leaders of each subgroup and the head of the evaluation team, assessed the technical solutions against the work demands of AEW&C, as specified in the work domain model. This process may be characterized as involving five main steps.

In the first step, the subgroups considered the technical solutions of each design concept, which could be mapped onto the physical objects and object-related processes levels of the model, with respect to how well they satisfied the purpose-related functions. To illustrate this step, Figure 9.4 presents an evaluation of a hypothetical design concept, Design Concept A.[†] Along the top of the figure are four purpose-related functions, which are the same as those in Figure 9.2. The headings down the left indicate that the figure reports three

---

[*] See Chapter 5 for a description of instantiations of structural means–ends relations.
[†] Information about the actual design concepts are not reported here for commercial and security reasons.

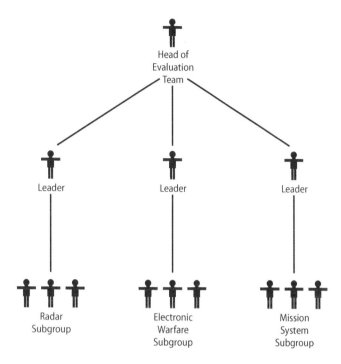

FIGURE 9.3
Team structure for the technical evaluation; only some of the subgroups are shown.

subgroups' assessments of Design Concept A, specifically those of the radar, electronic warfare, and mission system subgroups. The assessments, within the body of the figure, describe the effects of the respective technical solutions on the purpose-related functions.

Following the subgroups' assessments, the rest of the evaluation was performed by the panel. Specifically, for each design concept, the panel evaluated the combined impact of all of the technical solutions on every purpose-related function. Examples of such judgments for Design Concept A, based on the assessments of three subgroups, are shown at the bottom of Figure 9.4.

In the second step, the panel considered the effects that the findings at the purpose-related functions level would have on the value and priority measures. This step may be conceptualized with a representation similar to that in Figure 9.4. The value and priority measures would be listed at the top of the figure. The panel's findings for each purpose-related function would be reproduced down the left. (For Design Concept A, these findings would be obtained from the last row of Figure 9.4.) Against the findings for each purpose-related function, then, would be a description of their implications for each of the value and priority measures. The final row would show, for each value and priority measure, the implications of the findings across the set of purpose-related functions.

The third step involved extending the consequences of the technical solutions at the value and priority measures level to the functional purposes. This process, too, may be conceptualized with a representation analogous to those described previously.

In the fourth step, the panel compared the three design concepts in terms of how well their technical solutions satisfied the purpose-related functions, value and priority measures, and functional purposes. A comparison of three hypothetical design concepts, Design Concepts A, B, and C, against the aforementioned purpose-related functions is

Continued

	Detect, track, and associate	Evaluate tactical situation	Communicate	Implement protective measures
Radar Assessment	The radar's detection range of 600 nautical miles will enable earlier detection, tracking, and association of contacts than the original requirement	The radar's detection range of 600 nautical miles will allow earlier recognition of the emerging tactical situation than the original requirement	Communications to friendly assets will include information on tracks 600 nautical miles from the AEW&C system. The interception of these data communications by adversaries will reveal the AEW&C system's radar range, which breaches security requirements	The radar's detection range of 600 nautical miles will allow threats to be detected sooner and while they are further away from the AEW&C system, which means that operators will have more time to respond to threats
Electronic Warfare Assessment	The electronic support system detects fewer emitters than the original requirement. This will reduce the AEW&C system's capacity for identifying unknown entities in the environment	The electronic support system detects fewer emitters than the original requirement. This may result in an incomplete tactical picture and an inadequate understanding of the tactical situation	–	The electronic support system detects fewer emitters than the original requirement, although it can detect emitters from the most likely threats to the AEW&C system. The design concept does not meet the requirement for flares to be fitted to the aircraft. At altitude, the aircraft can be protected by positioning it outside the range of infrared missiles. During takeoff and landing, the AEW&C system will be vulnerable to ground-based infrared missiles. The design concept provides signature suppression significantly in excess of the original requirement thereby greatly reducing the effective range of an adversary's infrared missiles

FIGURE 9.4

Evaluation of a hypothetical design concept, Design Concept A.

	Detect, track, and associate	Evaluate tactical situation	Communicate	Implement protective measures
Mission System Assessment	-	The altitude of tracks is displayed adjacent to each track using height bars. Thus altitude information will be available to operators in a visuospatial format, which should help them in developing situation awareness. The original requirement was for altitude information to be displayed numerically	The design concept offers spatial audio technology, which will assist operators in distinguishing messages from multiple channels. Accordingly, message comprehension will be improved and the chance of missing important communications reduced	-
Panel's Evaluation	The reduced coverage of the electronic support system will compromise the identification of unknown entities. However, the enhanced radar range offers a significant advantage in detecting the existence of incoming tracks earlier. The location and activity information provided by the radar is judged to be more valuable than the capacity of the electronic support system to identify a large number of emitters. Specifically, it is of greater advantage to know the existence of a potential threat earlier than to know the specific identity of a threat. Hence the design concept exceeds the requirement for this function	The long radar range will allow operators to monitor track behavior for longer, which increases their likelihood of inferring the intent of a track. This will compensate for the deficiencies of the electronic support system. In addition, the visuospatial depiction of altitude may help operators in developing and maintaining situation awareness. Thus, overall, the design concept exceeds the requirement for this function	Protecting knowledge of the radar's range is a security imperative. This means that the AEW&C system must choose not to data link with friendly assets, which will significantly reduce its capability. Alternatively, the security of data links must be enhanced, which introduces difficult and expensive interoperability challenges. Thus, although it has been shown that spatial audio improves comprehension of voice communications, the design concept does not meet the requirement for this function	Earlier detection of threats allows timely evasive action and enhances the AEW&C system's capacity to remain outside the range of infrared missiles at altitude. However, the platform is still vulnerable to shoulder-launched infrared weapons during takeoff and landing. Although infrared signature suppression reduces the zone of vulnerability to some extent, flares would be of greater protective benefit. Thus the design concept does not meet the requirement for this function

FIGURE 9.4 (*Continued*)
Evaluation of a hypothetical design concept, Design Concept A.

presented in Figure 9.5. Against each design concept, the prior findings of the combined impact of the technical solutions on the purpose-related functions are reproduced. (For Design Concept A, these findings are taken from Figure 9.4.) The ranking of the design concepts for each purpose-related function is described at the bottom of the figure. Similar representations may be created to grade the design concepts with regard to how well their technical solutions fulfill the value and priority measures and functional purposes.

The final step involved establishing the overall ranking of the three design concepts across all of the purpose-related functions, value and priority measures, and functional purposes. The outcome was then justified to a board of senior defense personnel. It should be acknowledged that the panel's recommendations to the board may have been contingent on factors not embedded in the AEW&C model. For instance, faced with the hypothetical situation in Figure 9.5, the panel may have ranked Design Concept A last because it breaches critical security requirements. However, they may have advised that if the issues with this design concept's capability for data linking with friendly assets could be resolved (for example, by enhancing the security of the data links while maintaining interoperability requirements), it should be ranked first because of the significant benefits that would be conferred by a radar with a long detection range.

Two other points about the process of employing the AEW&C model for evaluation are worth noting. The first is that the technical solutions of the design concepts were assessed against all of the work demands of AEW&C, irrespective of the structural means–ends relations depicted in the model. The rationale for this approach was that the model displays instantiations of means–ends relations rather than the full set of possible means–ends relations, as described in the preceding section. Therefore, it was plausible that other structural means–ends relations might have been revealed during the evaluation process.

Finally, the resources made available to the evaluation team during this process included the graphical abstraction hierarchy and the tabular abstraction–decomposition space of AEW&C, as well as the glossary. The graphical abstraction hierarchy portrayed the structural means–ends relations between the constraints in the model, whereas the tabular abstraction–decomposition space showed the constraints at lower levels of detail or decomposition. The glossary served the purpose of describing the constraints and structural means–ends relations in more depth. The evaluation team was also provided with electronic templates for documenting its assessments of the design concepts.

Impact

One way of establishing the usefulness of work domain analysis for evaluating design concepts for complex sociotechnical systems is by assessing its impact or ability to influence practice. In the case of AEW&C, two main demonstrations of impact were apparent.

The first was a decision taken by the AEW&C System Program Office, the main defense unit in charge of the procurement of this system, to adopt work domain analysis as an additional approach for evaluation. Initially, only two approaches were planned: the technical and operational evaluations. Work domain analysis was viewed as offering a complementary perspective of the three design concepts, which would be beneficial for mitigating risk in selecting the best design for AEW&C.

Specifically, the AEW&C work domain model was seen to provide a common set of criteria for assessing the multiple, interacting solutions of each design concept and for comparing the different designs. Moreover, these criteria were work demands of the system that were relevant to a broad range of situations, including novel or unanticipated events. Consequently, this approach promoted an understanding of the extent to which the set of

Continued

	Detect, track, and associate	Evaluate tactical situation	Communicate	Implement protective measures
Design Concept A	The reduced coverage of the electronic support system will compromise the identification of unknown entities. However, the enhanced radar range offers a significant advantage in detecting the existence of incoming tracks earlier. The location and activity information provided by the radar is judged to be more valuable than the capacity of the electronic support system to identify a large number of emitters. Specifically, it is of greater advantage to know the existence of a potential threat earlier than to know the specific identity of a threat. Hence the design concept exceeds the requirement for this function	The long radar range will allow operators to monitor track behavior for longer, which increases their likelihood of inferring the intent of a track. This will compensate for the deficiencies of the electronic support system. In addition, the visuospatial depiction of altitude may help operators in developing and maintaining situation awareness. Thus, overall, the design concept exceeds the requirement for this function	Protecting knowledge of the radar's range is a security imperative. This means that the AEW&C system must choose not to data link with friendly assets, which will significantly reduce its capability. Alternatively, the security of data links must be enhanced, which introduces difficult and expensive interoperability challenges. Thus, although it has been shown that spatial audio improves comprehension of voice communications, the design concept does not meet the requirement for this function	Earlier detection of threats allows timely evasive action and enhances the AEW&C system's capacity to remain outside the range of infrared missiles at altitude. However, the platform is still vulnerable to shoulder-launched infrared weapons during takeoff and landing. Although infrared signature suppression reduces the zone of vulnerability to some extent, flares would be of greater protective benefit. Thus the design concept does not meet the requirement for this function
Design Concept B	The mission computer has a maximum storage limit of 200 tracks, which will result in a large number of entities in the environment remaining untracked. Moreover, the radar has a slow update rate (10 seconds for scan only, 15 seconds with tracking), which means that it is likely to provide inaccurate position information for tracks. Hence the design concept does not meet the requirement for this function	The greater display size will allow operators to have more information on the screen simultaneously. However, entities of interest may not be tracked because the mission computer only stores a maximum of 200 tracks, which will compromise operators' evaluation of the tactical situation. Moreover, the slow radar update rate will decrease operators' ability to anticipate track maneuvers. Hence the design concept does not meet the requirement for this function	The greater number of voice channels will reduce operators' workload in managing their voice communications. Thus the design concept exceeds the requirement for this function	The faster speed at which the platform can travel will give the AEW&C system more time to continue monitoring activity in the area of interest before it needs to take evasive action. Thus the design concept exceeds the requirement for this function

FIGURE 9.5
Comparison of three hypothetical design concepts, Design Concepts A, B, and C.

	Detect, track, and associate	Evaluate tactical situation	Communicate	Implement protective measures
Design Concept C	The mission system has advanced algorithms for associating tracks, which offers the potential for a greater number of track associations. Thus the design concept exceeds the requirement for this function	Information about the height of tracks can only be accessed via a third level of menu. Moreover, the system does not have the capability to display track history. Both of these features will have a negative effect on operators' evaluation of the tactical situation. Thus the design concept does not meet the requirement for this function	The design concept offers special processing features to improve the clarity of voice communications. Moreover, the design concept offers enhanced transmit and receive features, which will allow over-the-horizon communications. Thus the design concept exceeds the requirement for this function	The slower speed of the platform impairs its capacity to escape from threats. However, the design concept offers over-the-horizon communications, which will enable the platform to be positioned further from the battlefield. Thus the design concept satisfies the requirement for this function
Panel's Ranking of Design Concepts	The ranking of design concepts for this function is: *A, C, B. A* provides a significant tactical advantage by allowing earlier detection of incoming tracks. *C* associates pairs of tracks to which operators have not attended. However, such tracks are likely to be outside the area of significance and, therefore, of limited tactical interest. *B* is significantly disadvantaged by the limited track storage capacity of its mission computer and the slow update rate of its radar	The ranking of design concepts for this function is: *A, B, C. A* offers significant advantages in giving operators more time to monitor track behavior and infer the intent of tracks. *B* has the advantage of allowing operators to display more information on a screen. However, this advantage is outweighed by the limited track storage capacity of its mission computer and its slow radar update rate. *C* is significantly disadvantaged because operators will have no information about the history of tracks and height information is difficult to obtain	The ranking of concepts for this function is: *C, B, A. C* offers significant advantages as over-the-horizon communications will allow the AEW&C system to communicate with entities that would otherwise be out of range, and to act as a relay station for friendly entities that are beyond communication range with their base. *B* offers a small advantage in reducing the communications management workload, although this workload is not expected to be excessive. *A* is significantly disadvantaged by being unable to transmit on data links without breaching security requirements	The ranking of concepts for this function is: *B, C, A. B* has a significant speed advantage, which will allow the platform to remain on station for longer and to evade threats more successfully. *C* has the advantage of over-the-horizon communications, which implies that the platform can be located further from battle. However, its radar range is standard so the platform will need to remain forward to retain radar coverage of the battle. Its slower speed also inhibits its capacity to evade threats. *A* has a significant disadvantage in not having flares, which makes the platform vulnerable during takeoff and landing

FIGURE 9.5 (*Continued*)
Comparison of three hypothetical design concepts, Design Concepts A, B, and C.

technical solutions comprising each design concept would support the crew in dealing with the work demands of a variety of events, whether anticipated or unforeseen.

The AEW&C System Program Office arrived at the decision to adopt work domain analysis for evaluation following several presentations describing the approach and a demonstration of its potential benefits. The demonstration required a preliminary model of AEW&C to be constructed, which was based primarily on information from documents. This representation was used to provide concrete illustrations of how an evaluation using work domain analysis might proceed, as well as examples of the nature of the insights this approach could offer.

The second demonstration of impact was that the Australian government considered the results of work domain analysis in deciding which industry manufacturer to contract for the procurement of AEW&C. The selected manufacturer was, in fact, the one identified by this approach as providing the best design concept. The operational evaluation produced the same result. In a brief to the Deputy Secretary of the Australian Defence Organisation, the AEW&C System Program Office cited work domain analysis for its usefulness to the acquisition of this system.

Uniqueness

Another way in which the usefulness of this technique for evaluating design concepts can be established is by considering its uniqueness, or distinctiveness, relative to standard techniques used in industry. The technical and operational evaluations conducted for AEW&C may be regarded as standard approaches as they are commonly employed for the procurement of military systems (Charlton and O'Brien, 1996; Department of Defence, 1995, 1999; Gabb and Henderson, 1995, 1996; Malone, 1996; O'Brien, 1996). While there may be variations in how these approaches are implemented on different projects, their fundamental elements are the same.

As noted previously, a technical evaluation involves examining the technical solutions of a design concept against prespecified functional requirements. The example given for AEW&C was that the aircraft's radar must have the capacity to sense entities in the environment at a distance of 300 nautical miles. Typically, the technical solutions are assessed as complying with, surpassing, or failing to meet the requirements. This approach allows evaluators to develop an in-depth appreciation of the technical strengths or limitations of the design concept in relation to the requirements. However, it does not offer a systematic framework for establishing how effective a sizable set of interdependent technical solutions would be in fulfilling the system's work demands.

This requirement is addressed partially by the operational evaluation, which is concerned with establishing how well the set of technical solutions would fare in contending with the work demands of particular scenarios. In the case of AEW&C, this was achieved by testing the performance of a computer-based model of aspects of the technical solutions of a design concept in six scenarios using Monte Carlo simulation. With this approach, evaluators can gain detailed insight into the efficacy of a design concept with respect to the work demands of various scenarios. However, for practical reasons, the evaluation is normally limited to a small number of scenarios relative to the total set of possibilities, as was evident with AEW&C. Moreover, the scenarios tend to take the form of sequences of events or tasks that are known or can be predicted by subject matter experts. As a result, the operational evaluation does not promote an appreciation of how well a design concept would function in relation to the work demands of a wide range of situations, including unexpected events.

In the case of AEW&C, this requirement was met by work domain analysis. Specifically, the multiple, interacting technical solutions of a design concept were assessed against a set of constraints. These constraints, which specified the functional purposes, value and priority measures, and purpose-related functions that must be accomplished with a given set of physical resources, are event independent. Accordingly, the evaluation led to an understanding of how well the proposed design would support workers in meeting the work demands of a variety of situations, including those that are unforeseen. On the basis of this case study, it is apparent that work domain analysis complements the technical and operational approaches and makes a unique contribution to fulfilling the special requirements that complex sociotechnical systems pose for evaluation.

Feasibility

The feasibility of work domain analysis for evaluating design concepts for complex sociotechnical systems may be ascertained on the basis of whether it can be accomplished within the schedule, personnel, and financial restrictions of a project. In the case of AEW&C, the application was achieved within these limits. The schedule and personnel restrictions were described previously. The cost of the application was composed mainly of the remunerations provided to the analysts, subject matter experts, and evaluators. Although I did not keep exact records of these and other more minor expenses associated with the project, the outlay was no more than Aus$500,000 (Naikar, 2006c) and, in truth, was probably much less. The important point is that the outlay was relatively insignificant compared with the total cost of the acquisition of AEW&C, which was estimated at approximately Aus$3 billion.

Limitations

One limitation of this case study is that it does not provide empirical evidence that work domain analysis leads to the selection of the best design concept for a proposed system. It also does not demonstrate empirically that this form of analysis is more effective than standard techniques for evaluation. The resources necessary for conducting such studies in industrial settings are prohibitive.

To elaborate, one way to establish whether work domain analysis identified the best design concept for AEW&C would be to build three aircraft, one based on each of the three concepts, and to test the performance of these aircraft in a range of scenarios. Irrespective of the theoretical limitations of this approach, this option is highly impractical. Another approach would be to develop computer-based models of the design concepts and to test the performance of these models in a variety of scenarios using Monte Carlo simulation. Normally, this option, too, would be impractical. In the case of AEW&C, it was implemented to a degree in the form of the operational evaluation. The aim of the operational evaluation was not to validate the results obtained with work domain analysis but to offer a complementary perspective of the design concepts. Although both techniques identified the same design as the best for AEW&C, the computer-based models did not represent the full set of technical solutions of each concept. Hence, it is difficult to conclude with certainty that the results of the operational evaluation demonstrate that the design identified by work domain analysis was the best for AEW&C.

Such challenges with the empirical validation of techniques in industrial settings are not limited to work domain analysis. For this reason, techniques applied in industry are rarely evaluated formally (Czaja, 1997). Instead, the principle criterion against which they are gauged is usefulness (Vicente, 1999; Whitefield, Wilson, and Dowell, 1991). In this case, the usefulness of work domain analysis for evaluating design concepts for AEW&C was examined in terms of impact and uniqueness. Both criteria were satisfied on this project. In addition, the feasibility of this approach was demonstrated. Other case studies of the application of work domain analysis to this industrial problem would be beneficial.

Summary

The problem of evaluating alternative design concepts for complex sociotechnical systems deserves attention because, following the selection of a particular concept, only minor changes to the system design—those that are not inconsistent with the overarching concept—are practicable as a rule. Pinpointing the best design concept, therefore, is critical. In this chapter, it was proposed that work domain analysis is well suited for this application because it meets the special requirements that complex sociotechnical systems pose for evaluation. A case study was presented that showed how this approach was applied to AEW&C, a system under procurement by the Australian Defence Organisation. In particular, the process of performing the analysis of AEW&C was explained, in light of the analytic themes in Chapter 8, as well as the procedure for using the resulting model to evaluate the design concepts. On the basis of this case study, the usefulness and feasibility of work domain analysis for this novel application, which goes beyond designing ecological interfaces, was established.

10

Team Design

OVERVIEW Having described in the preceding chapter the application of work domain analysis to evaluate design concepts for complex sociotechnical systems, in this chapter I focus on the problem of developing team designs for such systems. More specifically, the problem of designing teams for future, first-of-a-kind systems, which poses significant challenges for work analysis, is considered. After establishing that cognitive work analysis is capable of surmounting these challenges, a detailed case study of this approach is presented. This case study shows how work domain analysis, together with control task analysis, was used to design a team for the Airborne Early Warning and Control system, which was also the subject of the preceding case study. Consistent with the aim of demonstrating the guidelines for work domain analysis in Chapter 8, the revision of the previous work domain model of this system is described in terms of those analytic themes. The chapter concludes by examining the usefulness and feasibility of cognitive work analysis for this application, as well as some limitations of this case study.

Context

The problem of team design has received surprisingly little attention from researchers given its significance for creating effective systems. Perhaps one reason for this apparent oversight is, in fact, the intractability of the problem. The challenge is even greater when one is faced with the task of designing teams for future, first-of-a-kind systems. Future systems are proposed systems that do not yet exist or have not yet been built. First-of-a-kind systems are those with no close existing analogues (Roth and Mumaw, 1995). These systems are considered revolutionary in their design, generally because they are marked by radical breakthroughs in technology.

Designing a team for a future, first-of-a-kind system when it is at the earliest stages of development is of the utmost importance. As the system is still at the conceptual phase at that time, the team design may be customized to its work demands. Later during development, the technical solutions are finalized, making it more likely that the team design will be driven by the technology. It is better to specify the team design early so that the technical solutions may be engineered in accordance with this design. While it is possible that the preferred team design may need to be modified, perhaps in light of financial considerations or technological obstacles, the focus is more likely to remain on finding ways to support this team design, and hence the fundamental work demands of the system.

Rationale

Designing teams for future, first-of-a-kind systems requires knowledge of their work demands. This information can be used to settle a range of questions about the most appropriate team design for a system, including the size of the team, its structure, the roles or responsibilities of team members, and their skills or training. On the whole, the intention in addressing these matters is to create a team design that best fulfills the system's work demands.

For several reasons, designing teams for future, first-of-a-kind systems poses special challenges for work analysis. As these systems are not yet functioning, their work demands cannot be established by studying or observing how work is performed in them. In addition, as these systems have no close existing analogues, their work demands cannot be inferred by investigating how work is carried out in current-generation systems. Another issue is that the work demands of these systems cannot be specified in full as a set of task sequences or procedures, given that the details of their technical solutions are still largely undefined, that workers often develop novel work practices as they gain experience with a new system, and that workers may have to deal with unanticipated events for which task sequences or procedures cannot be specified in advance.

Cognitive work analysis is well suited to the problem of designing teams for future, first-of-a-kind systems (Naikar et al., 2003). As explained in Chapter 1, this framework offers a formative approach to work analysis, which focuses on defining the constraints on actors' behavior. Specifying the constraints—or limits—that must be respected by actors does not require a functioning system. This means that, with this approach, it is possible to establish the work demands of future, first-of-a-kind systems without having to rely on examining how work is done in either these or current-generation systems. Furthermore, the constraints or limits on actors can accommodate many different trajectories of behavior, including task sequences or procedures that are difficult to specify a priori. In other words, these constraints are relevant irrespective of the technical solutions that are developed, the work practices that are invented, or the task sequences or procedures that are required for handling unforeseen events.

The strategy proposed here for using cognitive work analysis to design teams for future, first-of-a-kind systems has three main steps, which are presented in an idealized order. In this section, a brief overview of these steps is provided, emphasizing the rationale behind them. Details about how these steps may be implemented are offered in the next section in the context of a case study.

The first step is to perform a work domain analysis of the system with the aim of defining event-independent work demands. This framework provides a basis for evaluating alternative team concepts for the system with respect to how well they satisfy its work demands in a broad range of situations, including those that have not been anticipated.

The second step involves conducting a control task analysis to identify the system's work demands in recurring classes of situations. These work demands may be described as a set of work situations or work functions using the contextual activity template.* With this framework, the alternative team concepts may be characterized in terms of how they can manage the work demands of recurring classes of situations.

In the third step, the work domain analysis and the control task analysis from the preceding stages are used to develop a team design. Specifically, a tabletop analysis is carried out in

---

* See Figure 1.5a for an illustration of the contextual activity template.

which the alternative team concepts for the system are examined against its work demands. Tabletop analysis is well suited to this activity because its exploratory format allows subject matter experts and analysts to investigate the strengths and limitations of the alternative team concepts in relation to the system's work demands, rather than requiring that they understand these issues in advance.* The main output of this step is a set of requirements for a team design and, ultimately, a team design that satisfies those requirements.

Case Study

The case study reported here demonstrates how cognitive work analysis was applied to design a team for the Airborne Early Warning and Control (AEW&C) system which, at the time, was under procurement by the Australian Defence Organisation. A general description of this system was provided in the preceding chapter, as it was also the subject of that case study. AEW&C was considered to be a first-of-a-kind system because, despite the fact that its roles would be comparable to those of the Airborne Warning and Control System (AWACS) and E2C, it would have vastly different technology. In particular, AEW&C would use digital technology, whereas the then current-generation systems used analog technology. Digital technology would provide much greater operational flexibility. For example, the AEW&C aircraft's proposed radar could be stopped, started, and made to 'stare' at targets of interest at will, unlike the AWACS and E2C aircraft's radars, which are bound to a fixed number of revolutions per minute. AEW&C would also have significantly greater computing power, higher levels of data fusion, and more automation, including automatic radar tracking, radar operation, and situation and threat assessment. Consequently, the team design for AEW&C could not be based simply on those of AWACS or E2C.

Work Domain Analysis

As mentioned earlier, the first step of the strategy for team design is to perform a work domain analysis. This section explains how the original work domain model of AEW&C, which had been created to evaluate design concepts for this system, as described in Chapter 9, was revised for this application. As demonstrated later in this section, the main reason for the revision was not to tailor the model to the problem of designing a team for AEW&C as the model was found to be relevant to this problem. Instead, the model was amended in light of Vicente's (1999) views of work domain analysis, which were published after it had been produced.

The revision of the model is described in accordance with the analytic themes in Chapter 8. As was the case with the original analysis, these themes had not been formalized when the model was revised. However, as the process of modifying the model informed the themes that were developed, it can be readily conceptualized in those terms. Readers who are not interested in the details of the revision may proceed to the next section on the control task analysis of AEW&C.

---

* Tabletop analysis is explained in more depth in Theme 6 of Chapter 8.

Theme 1: Purpose of the Analysis

The objective of the revised model was to assist the Australian Defence Organisation with developing a team design for AEW&C, specifically for the aircraft. The strategy for how the model would be employed to achieve this aim was outlined in the preceding section.

Theme 2: Project Restrictions

The restrictions on the revision of the work domain model itself are difficult to isolate. Therefore, under this theme, the key restrictions on the entire application of cognitive work analysis to design a team for AEW&C are noted. The consequences of these restrictions for the revision will become apparent in the discussion of the succeeding themes.

The time available for this application was approximately six months. Three researchers from the Defence Science and Technology Organisation, myself included, were available for performing this work. Over the six-month period, I could allocate all of my time to this project, the second researcher could contribute approximately half of his time, and the third researcher could contribute approximately a third of his time. While I had conducted two work domain analyses previously, including the original analysis of AEW&C, the other researchers did not have any experience with this technique. None of us had experience in carrying out control task analysis or tabletop analysis for team design.

A military operator who had retired recently was contracted as a dedicated subject matter expert. This operator was involved in the original analysis of AEW&C.

Theme 3: Boundaries of the Analysis

The focus system for the revised model was the same as that for the original analysis. That is, the organizational entity delineating the boundaries was the Australian Defence Organisation's future AEW&C system, the physical entity was a single aircraft, and the focus problem was the development of a tactical picture and the coordination of defense assets in an allocated area, which was portrayed from the perspective of actors operating the aircraft.

As before, the boundaries excluded such components as ground-based mission planning and training facilities, as the objective was to create a team design for the aircraft. The focus problem was decided in view of the fact that the team design of the mission crew was judged to be of higher risk to the efficacy of AEW&C than that of the flight crew. As highlighted in Chapter 9, many of the technical solutions of the aircraft underpinning the focus problem were radically different in their design from those for AWACS and E2C. Therefore, the team design of the mission crew, whose primary concerns would be the work demands associated with this problem, was expected to be significantly different from those for current-generation systems. In contrast, the technical solutions relevant to the problem of flying the aircraft were similar to those for AWACS and E2C, not to mention a range of other military and commercial aircraft. Accordingly, the team design of the flight crew, whose main responsibilities would be connected with this problem, was expected to be similar to those for current-generation systems. Despite this, the team design of the flight crew was still taken into account in the evaluation of the alternative team concepts for AEW&C with the work domain model, although it was considered principally in relation to its adequacy in supporting the work demands linked with the focus problem.

Theme 4: Multiple Models

As was the case with the original analysis, multiple models could have been constructed to portray the focus problem separately from the perspectives of the flight crew and the mission crew. Detailed representations of the work demands of the two groups of stakeholders in relation to this problem could have facilitated a thorough assessment of the suitability of the alternative team concepts for supporting either set of work demands. Because of the project restrictions, a single model was retained for this application. Nevertheless, this representation covered the perspectives of both crews by depicting the focus problem from the viewpoint of actors operating the aircraft. As a result, it still led to an appreciation of the viability of the alternative team concepts for supporting the work demands of the two crews, although not in as much detail.

Similarly, as discussed for the original analysis, multiple models could have been created to highlight the distinct facets of the focus problem, specifically risk assessment and risk management. With this approach, a detailed understanding of the suitability of the alternative team concepts for supporting the work demands associated with each facet could have been obtained. However, this approach would have been difficult to achieve within the project restrictions. Therefore, once again, a single model was retained. Instead of separating the representation of risks from that of resources for managing those risks, this model encompassed both risks and resources, making the interactions between these elements more apparent. Consequently, an appreciation of the effectiveness of the alternative team concepts in relation to both risk assessment and risk management was still gained, although compared with a multiple-model approach, the assessment was probably less detailed and less explicit. On the other hand, the single model most likely promoted a better understanding of the suitability of the alternative team concepts for supporting the work demands arising from the interactions between risks and resources. While the ideal solution may have been to implement both approaches, this was not possible within the project restrictions.

Given that the outcomes of the two preceding decisions, about where to place the boundaries of the analysis and whether to create multiple models, were the same as those for the earlier analysis, the original model was also relevant for team design. However, as mentioned before, the model was revised to take into account Vicente's (1999) interpretation of work domain analysis, which was published after the model had been developed. In particular, Vicente stressed that a work domain model should focus on the objects of action, and that these properties should be labeled with nouns, not verbs.* These ideas prompted changes in how constraints in the model were represented. For the sake of completeness, these modifications are described within the context of the remaining themes. Most of these themes are addressed only briefly because many of the remaining details of this analysis are similar to those for the analysis in Chapter 9.

Theme 5: Causal–Intentional Continuum

The focus system was located toward the right of the middle of the causal–intentional continuum, which means that it was categorized as a loosely coupled intentional system. Actors' behavior would be governed principally by intentional rather than causal con-

---

* Vicente's ideas are discussed in more detail in Chapter 7.

straints. Accordingly, the model of the focus system was dominated by intentional constraints at higher levels of abstraction and causal constraints at lower levels.

Theme 6: Sources of Information

The primary information source for the revised model was the original model, including the glossary. The data obtained from documents, field observations, and subject matter experts for the original analysis were also reassessed for the revision.

Theme 7: Content of the Abstraction–Decomposition Space

The content of the original model was revised by examining the constraints and reworking their descriptions to emphasize objects of action. In addition, the constraints were labeled with nouns. Figure 10.1 depicts a sample of the revised model; the labels of some of the constraints have been edited slightly to make the sample comprehensible. Although this model appears markedly different from the original, there was only one fundamental difference between the two. Specifically, within a level of abstraction, several constraints in the original model were folded into a single constraint in the revised model. For example, at the purpose-related functions level, constraints such as *detect, track, and associate* and *evaluate tactical situation* were combined into the description *portrayal of tactical situation*. The rationale for this modification was that both of the original purpose-related functions are concerned with the tactical picture, the former with its development and the latter with its evaluation. Hence, the object of action for both purpose-related functions is the tactical picture, which was represented in the revised model as *portrayal of tactical situation*. This label was preferred to the label 'tactical picture' because, although the latter

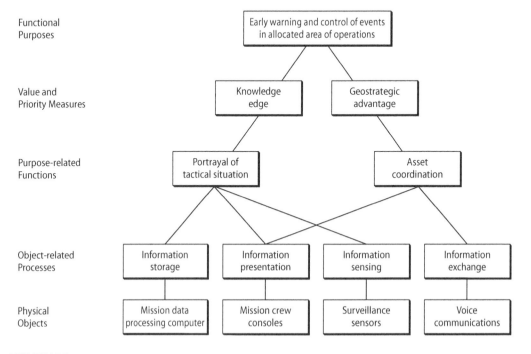

FIGURE 10.1
Sample of the revised AEW&C model.

is more readily viewed as describing an object of action, it does not seem to denote a function or, more specifically, a purpose-related function. In addition, the preferred label was chosen over the label 'portray tactical situation' so that this constraint was represented with a noun.

The merging of constraints in the revised model occurred mainly at the top four levels of abstraction. The constraints at the lowest level, physical objects, were not affected in this way because they already emphasized objects of action. Nevertheless, the descriptions at this level were also adjusted, mainly to ensure consistency with the granularity of the descriptions at the higher levels.

The revised model, then, was fundamentally the same as the original model. Both representations captured the same constraints at each level of abstraction, albeit at different granularities; the constraints in the revised model were depicted at a coarser granularity. Furthermore, the constraints in the revised model were labeled with nouns, whereas those in the original model were labeled mainly with verbs. As it was possible to match the constraints in the two representations, the decomposition of the constraints in the original model was kept as a supplement to the revised model. For the same reason, the glossary that had been created for the original model was retained. The reuse of these aspects of the original model was also appropriate in light of the project restrictions.

Given the degree of similarity between the two models, it is worth reflecting on the value of revising the original representation to accommodate Vicente's (1999) ideas. By stating that the descriptions of constraints were modified to emphasize objects of action and to replace verbs in the labels with nouns, it is implied that the original descriptions signified actions. However, these descriptions were not intended to signify actions but affordances.* For instance, the purpose-related function *evaluate tactical situation* was not intended to denote the activity of evaluating the tactical situation but to capture the fact that the system affords the function of evaluation of the tactical situation. In the preceding sentence, the term *function* is used as a noun to indicate an affordance, not as a verb to indicate an activity. Moreover, the affordances portrayed in a model may be considered objects of action in the sense that actors within a system act on the affordances of their environment. Nevertheless, the original descriptions did tend to be misinterpreted as activity by various viewers of the model, perhaps because of the use of verbs in the labels. Therefore, revising the original model to substitute the verbs with nouns was worthwhile.

Theme 8: Validity of the Model

As was the case for the original model, the strategy for validating the revised model was to have it reviewed by subject matter experts. The review of the revised representation was less extensive than that conducted originally, especially with respect to the number of participants. However, the types of experts involved in the review and the form of the review were essentially the same. In the end, little refinement of the revised model was necessary.

Control Task Analysis

The second step of the strategy for team design involves conducting a control task analysis. This section provides a brief description of the analysis performed for AEW&C. Further details of this analysis are provided by Naikar et al. (2003, 2006).

---

* See Chapter 7 for a discussion of the representation of affordances in a work domain model.

Control task analysis involves defining a system's work demands in recurring classes of situations.* It may be viewed as comprising two main stages (Elix and Naikar, 2008a; Naikar et al., 2006; Rasmussen et al., 1994). First, the work demands of the system are characterized in terms of work situations or work functions. Following that, the control tasks that are necessary for each work situation or work function are specified. Only the first stage of control task analysis is utilized in this case study.

Depending on the system, the work demands identified by control task analysis may be in the form of a set of work situations, work functions, or a combination of both. Work situations are relevant when activity is structured according to time or space. Work functions, on the other hand, are appropriate when activity is organized on the basis of its functional content, irrespective of the time or space in which it occurs. In some systems, a combination of the two is most suitable for representing work demands. This is the case for systems in which activity within a work situation, or set of work situations, is further delineated in terms of its functional content. For such systems, the combinations of work situations and work functions that are possible may be depicted with a contextual activity template.

A contextual activity template for AEW&C is presented in Figure 10.2. The work situations, depicted along the horizontal axis, are the different phases of a mission. The work functions, shown in the circles, signify different types of problems. The boxes spanning the work functions indicate those work situations in which a work function *can* occur. The bars extending from the work functions indicate those work situations in which a work function *typically* occurs. The contextual activity template is supplemented with a glossary that describes each work function in more depth. For example, in the case of *manage asset disposition*, the glossary reveals that this work function concerns the problem of managing the number, types, and disposition (location, weapon status, fuel status, tasking) of assets to achieve the aims of the mission under changing tactical and environmental conditions.

Team Design

The third step of the strategy for team design involves undertaking a tabletop analysis in which the work domain analysis and control task analysis from the preceding stages are used to develop a team design for the system. The tabletop analysis is composed of six substeps, which involve the following activities:

1. Defining alternative team concepts for the system
2. Designing scenarios that represent the types of events with which the system may have to contend
3. Exploring how the team concepts can manage the work demands posed by those scenarios
4. Characterizing the team concepts in terms of the work demands of recurring classes of situations, as specified by the control task analysis
5. Evaluating the team concepts against the work demands of a broad range of situations, as specified by the work domain analysis, and formulating requirements for a team design on this basis
6. Creating a team design that satisfies the requirements

---

* More information about control task analysis is provided in Chapter 1 and the Appendix.

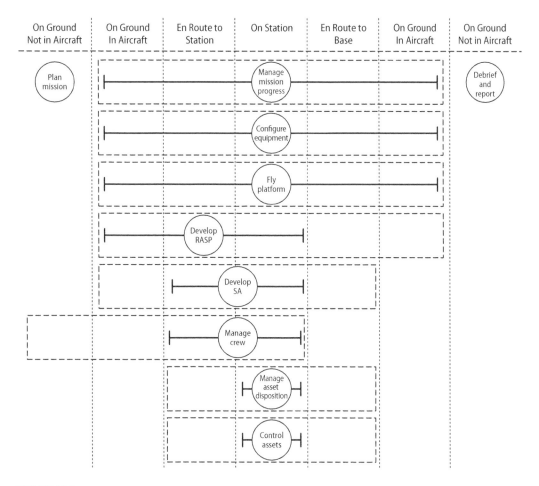

FIGURE 10.2
Contextual activity template for AEW&C. RASP, Recognised Air and Surface Picture; SA, situation awareness.

This section explains these substeps in more detail and describes how they were implemented for AEW&C.

Substep 1: Definition of Alternative Team Concepts

It is possible to specify alternative team concepts for the proposed system in terms of a set of variables (e.g., size of team, number of levels of hierarchy) and plausible values for each variable (e.g., two, three). As is demonstrated in this section, the different combinations of values across the set of variables identify the alternative team concepts. Not all of the team concepts need to be examined explicitly in the subsequent substeps; some may be disregarded on the strength of the results obtained with particular team concepts.

The team concepts for AEW&C were defined through discussions with subject matter experts, who had similar kinds of knowledge and experience to those who were consulted for the work domain analysis.[*] Figure 10.3 shows the team concepts for the mission crew. The variables underpinning these team concepts are the size of the team, the number of

---
[*] See Chapter 9 for details about the subject matter experts consulted for the work domain analysis.

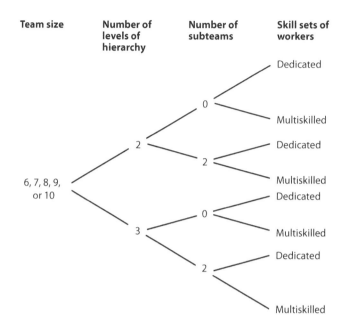

FIGURE 10.3
Alternative team concepts for the mission crew.

levels of hierarchy, the number of subteams, and the skill sets of workers. For the variable of team size, for instance, values between six and ten were considered plausible for AEW&C. Alternatively, for the variable of skill sets, both workers with dedicated roles or responsibilities and workers who were multiskilled were regarded as plausible.

In Figure 10.3, the different pathways—or combinations of values across the set of variables—specify the team concepts. For example, the top pathway stipulates a team concept with six people, two levels of hierarchy, no subteams, and dedicated workers. The next pathway identifies a team concept with six people, two levels of hierarchy, no subteams, and multiskilled workers.

Substep 2: Scenario Design

This substep involves designing scenarios that represent the types of circumstances the system may be tasked or confronted with handling. The purpose of these scenarios is to provide subject matter experts with a context for exploring possibilities for organizing work with the alternative team concepts in a manner that is consistent with the nature of their expertise. In its most basic form, scenario design entails specifying a series of critical events and establishing a timeline for those events.

The scenarios for AEW&C were constructed by the analysts and two subject matter experts. Both experts were military personnel who had operational experience in an Australian ground-based system with similar roles to AEW&C. They also had operational experience in comparable current-generation systems: one in AWACS and the other in E2C. In addition, the two experts possessed extensive knowledge of the proposed operational and system concept for AEW&C, having contributed to the development of its concept of operations when they were based at the AEW&C System Program Office, the main defense unit responsible for procuring this system.

Two scenarios were prepared specifically for the tabletop analysis. One scenario, which entailed conducting general surveillance activities in the northern regions of Australia, typified a routine mission for AEW&C. The other scenario, which involved supporting allied forces in battle, represented one of the more challenging circumstances in which AEW&C may be required to participate. Another six scenarios that had been developed by operations analysts for evaluating the system design concepts for AEW&C, the application described in Chapter 9, were also available.

To construct the two scenarios, a series of critical events was specified for each. These events would require some response (e.g., actions, decisions) from members of the crew. A critical event may involve, for example, a strike package entering a no-fly zone or the hostile firing of friendly ships. As well as establishing a timeline for these critical events, paper snapshots of each event were developed. These were hand-drawn illustrations of each event shown against a backdrop of the relevant geographical area. Textual descriptions of the scenes were also included.

Substep 3: Tabletop Sessions

In this substep, a series of tabletop sessions is conducted in which subject matter experts explore how the alternative team concepts for the proposed system can manage the work demands of the scenarios prepared earlier. Specifically, in each session, experts are required to define the work demands generated by the critical events in a scenario and to examine how those work demands can be handled by one of the team concepts. The judgments elicited from experts during this process should resemble those they make in their actual occupations and, therefore, should be consistent with the nature of their expertise. The format of the sessions should be such that experts are not subjected to the time or other pressures present in real scenarios. This makes it practical for experts to contemplate a variety of possibilities for work allocation in a scenario with a given team concept, and to examine the strengths and limitations of those options in depth, before making a final decision about the most suitable option.

For AEW&C, the experts who participated in the tabletop sessions were the same two military personnel who were involved in designing the scenarios in the preceding substep. It was not necessary to have experts without prior knowledge of these scenarios because the aim was not to test the performance of either the experts or the team concepts in the scenarios. Instead, as indicated above, the objective was for the experts to explore the possibilities for work allocation in the scenarios with the alternative team concepts.

For each session, the analysts selected a scenario and a team concept for the experts to inspect. The experts first considered the roles and, therefore, the nature of the work demands that could be assigned to crew members, given the specified team concept. For instance, in a team concept with multiskilled workers, a person could be allocated the roles of both fighter controller and surveillance officer, whereas in a team concept with dedicated workers, a person could be given only one of those roles. Subsequently, the experts identified the work demands that would arise from the critical events in the scenario and investigated how those work demands could be handled by the team concept.

Detailed records were kept of each tabletop session. Specifically, representations were developed of the work demands that crew members were allocated as a function of the critical events. These representations were much like tables with the crew members shown in the rows, the critical events indicated in the columns, and the work demands that crew members were assigned portrayed in the cells. The analysts also kept written notes of each tabletop session. In particular, notes were taken of the different possibilities the experts considered for work allocation, the strengths and limitations they identified for each option, and the criteria they took into account in deciding which option to select.

Substep 4: Characterization of Team Concepts in Terms of Work Functions

This substep involves translating the work demands in the representations created earlier into work functions from the control task analysis. The purpose of this step is to characterize the team concepts in terms of how they can manage the work demands of recurring classes of situations, rather than the work demands of particular scenarios. This means that the observations from the tabletop sessions will extend to a larger range of scenarios. In addition, the alternative team concepts can be compared across different scenarios as the work demands of the scenarios are conveyed in the same terms. Another advantage is that the allocation of work to team members is expressed in the form of a set of constraints, which can accommodate a variety of task sequences or procedures, including those that cannot be specified a priori.

In the case of AEW&C, the process of transforming scenario-specific work demands into work functions was guided by the glossary for the control task analysis. A work demand that was concerned with 'directing a fighter aircraft X to conduct a sweep of a location Y,' for example, was converted into the work function of *control assets*. On the other hand, a work demand that involved 'changing the radar search pattern from mode X to mode Y' was translated into the work function of *configure equipment*.

The final characterizations of the team concepts in terms of work functions are too large to be displayed in full here. Therefore, Figure 10.4 presents sections of two of the characterizations. These sections portray two team concepts at the same point in one of the scenarios. The main difference between these team concepts is the size of the team; Figure 10.4a depicts a team concept with ten people, whereas Figure 10.4b depicts a team concept with six people. Otherwise, both team concepts are the same, having two levels of hierarchy, no subteams, and multiskilled workers.

As shown in the figure, the characterizations contained the roles of the crew members (e.g., mission commander, fighter controller) in the rows; the critical events (e.g., identifying tracks of interest, control of P-3C aircraft) in the columns; and the work functions (e.g., control assets, configure equipment) in the circles. The arrows denoted the coordination requirements for reorganizing responsibilities among the crew, which the experts considered in allocating work demands to the team. In Figure 10.4b, the unfilled rows signify members of the ten-person team who were not available in the six-person team. The circles in bold highlight the additional responsibilities of workers in the six-person team compared with the ten-person team.

Substep 5: Evaluation of Team Concepts and Formulation of Requirements for Team Design

In this substep, the work domain model constructed previously is used to evaluate the alternative team concepts and specify requirements for a team design. Specifically, the team concepts are evaluated on the basis of the impact of their respective distributions of work functions on the system's effectiveness in fulfilling its work demands in a broad range of situations, including unforeseen events. The aim of this process is not to establish whether the team concepts are effectual or ineffectual on the whole but, instead, to identify their strengths and limitations. Consequently, a set of requirements for a team design may be formulated that seeks to maximize the positive impact, or minimize the negative impact, of the team concepts on the system's effectiveness.

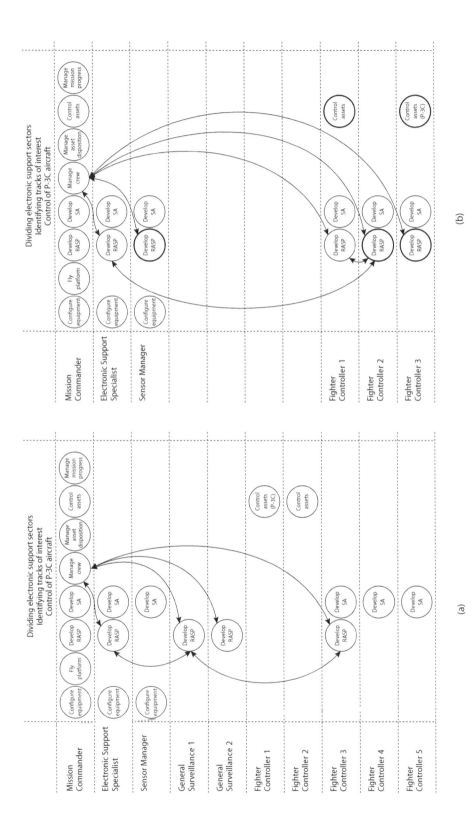

FIGURE 10.4

Sections of the characterizations of two team concepts at the same point in a scenario: (a) ten-person team; (b) six-person team. RASP, Recognised Air and Surface Picture; SA, situation awareness.

Figure 10.5 presents a sample of observations from this substep of the tabletop analysis for AEW&C. The scenario that resulted in these observations was an eight-hour mission with high-level conflict in which AEW&C was tasked with focal area surveillance of a no-fly zone of Country X and broad area surveillance of a maritime region. The observations relate to the following team concepts: a six-person team with two levels of hierarchy, no subteams, and dedicated workers; a six-person team with two levels of hierarchy, no subteams, and multiskilled workers; a ten-person team with two levels of hierarchy, no subteams, and dedicated workers; and a ten-person team with two levels of hierarchy, no subteams, and multiskilled workers. Thus, the main differences between these team concepts were the number of people in the team and whether the crew members were dedicated or multiskilled.

To evaluate the team concepts for AEW&C, the analysts first identified the major patterns that were evident in the distributions of work functions. In particular, different and recurring patterns in the characterizations were noted. In general, many of the patterns related to the number of crew members concerned with particular work functions, the number and combinations of work functions allocated to particular crew members, and the nature of the instances in which work functions were reorganized among the crew. Some examples of such patterns are given in Figure 10.5.

Following that, the analysts and the subject matter experts, who were the same two military personnel involved in the preceding substeps, assessed the patterns against the work demands of AEW&C, as specified in the work domain model. Specifically, the ramifications of the patterns for each of the system's functional purposes, value and priority measures, purpose-related functions, object-related processes, and physical objects were considered. Figure 10.5 provides some examples of such judgments.

Subsequently, the analysts formulated requirements for a team design for AEW&C, which were reviewed by the subject matter experts. As suggested, the requirements sought to enhance the positive impact, or reduce the negative impact, of the team concepts on the system's efficacy in meeting its work demands. Once again, some examples of such requirements are presented in Figure 10.5.

It is worth noting that, although some of the observations in Figure 10.5 may seem obvious in hindsight, not all of the findings of the tabletop analysis were anticipated by the analysts or the experts. In addition, while some of the observations were certainly predictable, the full implications of these findings for AEW&C had not been considered explicitly, or systematically, prior to the tabletop analysis. Moreover, the analysts and the experts found some of the observations surprising.

Substep 6: Development of Team Design

In the final substep, a team design that fulfills the requirements specified previously is created. As highlighted earlier, these requirements are formulated so that the positive aspects of the team concepts are enhanced and their negative aspects are reduced. The resulting team design, then, should capitalize on the strengths of the alternative team concepts while mitigating their limitations.

In the case of AEW&C, the analysts studied the requirements from the preceding substep and developed a suitable team design. This team design accommodated flexibility in the size of the team for the mission crew (either six or seven people) and in the number of levels of hierarchy (either two or three levels). In addition, it was specified that the team should not be decomposed into subteams, and that workers should be multiskilled. The team design for AEW&C, therefore, could be adapted to different situations.

Continued

Different and Recurring Patterns	Impact of Patterns on Work Demands	Requirements for Team Design
Members of the ten-person teams were responsible for fewer work functions than members of the six-person teams.	Members of the ten-person teams may have too few responsibilities. If workload is low, team members may become bored and vigilance may be poor (Huey and Wickens, 1993). Poor vigilance is most likely to lead to detriments in the performance of routine activities, such as general maintenance of the tactical picture. As a result, the purpose-related function of *portrayal of tactical situation* and the value and priority measure of *knowledge edge* may be compromised. In contrast, members of the six-person teams may have too many responsibilities. If workload is high, some tasks may be shed or prioritized over others (Huey and Wickens, 1993). Routine activities like general maintenance of the tactical picture are likely to be given lower priority than activities relating to coordinating and protecting valuable defense assets in a hostile airspace. Thus the purpose-related function of *portrayal of tactical situation* and the value and priority measure of *knowledge edge* may be compromised in the six-person teams as well.	In complex scenarios, a six-person team may be inadequate for the AEW&C system. However, a ten-person team may be unnecessarily large. A seven-person or eight-person team may be suitable.
In the teams with multiskilled workers, in which responsibilities could have been reallocated or reshuffled amongst team members, there were no instances of the reallocation of work functions.	There were no apparent benefits of teams with multiskilled workers on the work demands of the AEW&C system in these cases. The ten-person team had so much spare capacity that there was no need for responsibilities to be reorganized amongst team members. There was always a team member available with the capacity for dealing with a new work demand. In the six-person team, all of the team members were concerned with several work functions at all times. Hence there was little spare capacity for team members to reorganize their responsibilities in response to local contingencies.	A team size of seven or eight may be required for the AEW&C system to realize the benefits of a team with multiskilled workers.

FIGURE 10.5
Sample of observations for AEW&C.

Different and Recurring Patterns	Impact of Patterns on Work Demands	Requirements for Team Design
Of the six-person teams, at least one member of the team with dedicated workers was concerned solely with the general maintenance of the tactical picture throughout the scenario. In contrast, members of the multiskilled team who were responsible for general maintenance of the tactical picture were also controlling or managing assets at critical points in the scenario.	In the six-person team with dedicated workers, not all team members had the skills for controlling or managing assets. Therefore, at least one person was dedicated to the general maintenance of the tactical picture throughout the scenario. Conversely, in the six-person team with multiskilled workers, all team members who were concerned with the general maintenance of the tactical picture were also controlling or managing assets at critical points in the scenario. Given the challenges associated with coordinating and protecting valuable defense assets in a hostile airspace, the multiskilled team may find it more difficult than the dedicated team to satisfy the purpose-related function of *portrayal of tactical situation and the value and priority measure of knowledge edge.*	In complex scenarios, at least one member of a small team should be responsible primarily for the general maintenance of the tactical picture, even if team members are multiskilled.
Of the six-person teams, more members of the multiskilled team were responsible for controlling or managing assets compared with the team with dedicated workers.	In the six-person team with multiskilled workers, all team members had the skills for controlling or managing assets. In contrast, in the six-person team with dedicated workers, only three team members had the skills for controlling or managing assets. As the most significant work demands in the scenario were those associated with controlling or managing assets, responsibilities were distributed more evenly across members of the multiskilled team. Thus the team with dedicated workers may find it more difficult than the team with multiskilled workers to satisfy the purpose-related function of *asset coordination and the value and priority measure of geostrategic advantage.*	Members of small teams should be multiskilled to allow responsibilities to be distributed evenly across the team.
In all four team concepts, the mission commander had significant responsibilities for monitoring or coordinating crew and liaising with external agencies.	In all four team concepts, the mission commander's administrative or 'housekeeping' duties may be detrimental to his or her primary responsibility of understanding the emerging tactical situation and making effective tactical decisions. Hence the purpose-related functions of *portrayal of tactical situation and asset coordination and the value and priority measures of knowledge edge and geostrategic advantage may be compromised.*	For complex scenarios, a deputy mission commander, who assumes the role of supervisor or coordinator, may be beneficial.

FIGURE 10.5 (Continued)
Sample of observations for AEW&C.

Figure 10.6 shows an instantiation of this team design in which the role of deputy mission commander is implemented when the mission commander needs to be buffered from administrative duties. On missions, or segments of missions, when this buffer is not required, the person in this role can assume other responsibilities or serve as a spare crew member to enable the rotation of crew through positions thus allowing more or longer rest periods for members of the team. These configurations involve variations in team size, hierarchical structure, and roles or responsibilities.

Impact

The usefulness of cognitive work analysis for team design can be established on the basis of its impact or ability to influence practice. In the case of AEW&C, there were two clear demonstrations of impact.

One was that the resulting team design was adopted by the AEW&C System Program Office. This team design was judged by subject matter experts within this organization to be suitable for AEW&C and better than other team designs they had considered. The experts included the two military personnel who had participated in the tabletop analysis, other military personnel with similar backgrounds to them, and senior defense personnel with broader, strategic knowledge of AEW&C. As an empirical evaluation of the team design was not feasible, for reasons that are discussed later in this chapter, the AEW&C System Program Office implemented a number of strategies to manage the risk that the team design may, in actual fact, be unsatisfactory. In particular, each aircraft would be equipped with ten consoles in case a team size of seven for the mission crew was found to be insufficient. Furthermore, evaluations of the team design would be conducted as soon as mockups, prototypes, or simulations of the aircraft became available.

Another demonstration of impact was that, on the basis of the analyses carried out to develop a team design for AEW&C, modifications were made to the technical specifications for the aircraft. For example, the number of communication channels on the aircraft

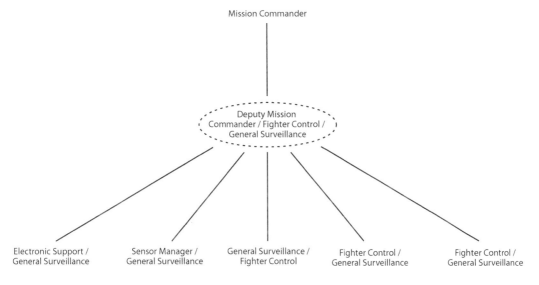

FIGURE 10.6
Instantiation of team design for AEW&C.

was increased. Such modifications, which were stipulated before the Australian government signed a contract with the industry manufacturer, were achieved without cost.

Uniqueness

The usefulness of cognitive work analysis for team design can also be confirmed on the basis of its uniqueness compared with other techniques more prevalent in industry. Conventional techniques span a variety of disciplines, including engineering (Davis and Wacker, 1982, 1987; Mundel, 1985; Niebel, 1988); social and organizational psychology (Hackman and Oldham, 1980; Medsker and Campion, 1997; Sundstrom, De Meuse, and Futrell, 1990); and engineering psychology (Dieterly, 1988; Dubrovsky and Piscoppel, 1991; Lehner, 1991). Although these techniques differ markedly in their theoretical orientations, they all rely on some form of work analysis as a means of arriving at a suitable team design for the system in question, that is, a team design that will fulfill the system's work demands effectively.

Conventional techniques have been useful for designing teams for existing systems (Medsker and Campion, 1997), but they cannot be readily applied to future, first-of-a-kind systems. Some of these techniques depend on descriptive methods of work analysis, which seek to understand how work *is done* in a system (Vicente, 1999). However, given that future, first-of-a-kind systems are not yet operational, it is not possible to investigate how work is conducted in these systems. Another problem is that, because these systems are revolutionary in their design, their work demands cannot be deduced on the basis of how work is accomplished in current-generation systems.

Other conventional techniques depend on normative methods of work analysis, which seek to specify how work *should be done* in a system in terms of a set of task sequences or procedures (Vicente, 1999). However, without the full details of their technical solutions, it is not possible to define the work demands of future, first-of-a-kind systems comprehensively in this form. Further issues are that workers tend to devise innovative ways of working as they become more practiced in operating a system, and that they may be faced with managing unforeseen situations, which also makes it unfeasible to specify the entire set of task sequences or procedures up front.

In contrast to conventional techniques, cognitive work analysis is ideal for designing teams for future, first-of-a-kind systems. Instead of focusing on how work is done or should be done in a system, this framework is concerned with defining the constraints on actors. As explained previously, the analysis of constraints does not depend on having a functioning system. Moreover, the constraints on actors are applicable regardless of their current or future work practices, the circumstances they encounter, or the details of the technical solutions available to them.

Cognitive work analysis made it possible to develop a team design for AEW&C, a future, first-of-a-kind system, that is customized to its work demands. If we had relied on descriptive methods, the team design could not have been established until the system became operational or, at best, until later during its development. As a result, the team design might have been constrained artificially by the system's technical solutions. Alternatively, the descriptive methods could have been applied to current-generation systems, such as AWACS or E2C, to infer the work demands of AEW&C. However, the resulting team design might not have taken full advantage of AEW&C's revolutionary technology, and it might have incorporated, inadvertently, unproductive work practices in current-generation systems.

On the other hand, if we had relied on normative techniques, the team design for AEW&C would have been based on a set of task sequences or procedures. However, given the difficulty of describing work demands fully in this form, this set would have been incomplete. As a result, the team design might not have been well suited to work demands that could not be defined a priori. By leading to team designs that satisfy a set of constraints, instead of a set of task sequences or procedures, cognitive work analysis promotes the development of flexible teams, which are capable of dealing with the work demands of a broad range of situations, including unanticipated events.

Feasibility

The feasibility of applying cognitive work analysis to the problem of team design may be established on the basis of whether this approach can be implemented within a project's schedule, personnel, and financial restrictions. This criterion was met in the case of AEW&C. The schedule and personnel limits on this project were outlined previously. The main financial cost was the wages of the analysts and the subject matter experts. However, as observed in relation to the preceding case study, the cost of this application was substantially less than the overall budget of the AEW&C acquisition.

Limitations

Arguably, one limitation of the proposed strategy for using cognitive work analysis to design teams for future, first-of-a-kind systems is that it relies on judgments by subject matter experts. Specifically, experts are required to judge how the work demands of various scenarios can be managed by the alternative team concepts, as well as what impact the different distributions of work functions across team members will have on the system's effectiveness. While it is difficult to be certain about the robustness of experts' judgments during a tabletop analysis, these appraisals resemble those they make in their real jobs and, therefore, do not exceed the bounds of their expertise. In any case, for future, first-of-a-kind systems that are still at the early stages of development, there appears to be no better alternative. When mockups, prototypes, or simulations of these systems become available, the performance of any team designs created for them should be evaluated.

Like the case study in the preceding chapter, some limitations of this case study are that it does not offer empirical evidence that cognitive work analysis leads to effective team designs for future, first-of-a-kind systems or to better team designs for such systems than those based on other techniques. However, such evidence is difficult, if not impossible, to obtain in industrial settings because of the resources that are required.

For example, in order to evaluate whether cognitive work analysis resulted in an effective team design for AEW&C, access to the real aircraft, or to mockups, prototypes, or simulations of relevant features of the aircraft, would be necessary. Such resources were not available when this team design was developed. Besides, even with access to these resources, there are other challenges associated with establishing whether cognitive work analysis led to a valid team design. The main difficulty is distinguishing the contribution of this approach to the efficacy of the team design from the part played by other features of the aircraft. For instance, if the team design performs well, it may be the case that its effectiveness is due to the aircraft's interface design, such that any team design would have performed successfully with that interface.

Similar challenges would be faced in carrying out empirical studies to determine the value of cognitive work analysis for team design relative to other techniques. In the case of AEW&C, it would have been necessary to develop additional team designs for this system using different techniques. Subsequently, the performance of the various team designs would have to be compared using the real aircraft or approximations of it. The resources necessary for performing this type of study were unavailable.

As highlighted in the context of the preceding case study, these kinds of difficulties with the empirical validation of techniques in industrial settings are not particular to cognitive work analysis. This is why techniques used in industry tend to be assessed against the standard of usefulness (Vicente, 1999; Whitefield et al., 1991) rather than being subjected to a formal evaluation (Czaja, 1997). In the present case, the usefulness of cognitive work analysis for designing a team for AEW&C was established on the basis of impact and uniqueness, and the feasibility of this approach was demonstrated as well. Further applications of this framework to the problem of team design are necessary to determine whether it is useful and feasible consistently.

Summary

This chapter was concerned with the challenging problem of designing teams for future, first-of-a-kind systems, especially during the initial stages of their development when the team designs can be tailored to the systems' work demands. Cognitive work analysis provides a powerful framework for addressing this problem because it focuses on the analysis of constraints. As a result, it relies neither on having a functioning system nor on knowing in full the particularities of the system's technical solutions, the current or future practices of workers, or the situations that workers will face. A case study was presented that demonstrated how work domain analysis and control task analysis were applied to design a team for AEW&C, a future, first-of-a-kind system intended for the Australian Defence Organisation. This case study not only provided another illustration of the guidelines for work domain analysis in Chapter 8 but also served the key purpose of establishing the usefulness and feasibility of cognitive work analysis for this novel application.

11

Training

OVERVIEW   In this chapter, the last of this book, one more application of work domain analysis is considered—that of establishing the training requirements of complex sociotechnical systems. First, a case is made for the value of work domain analysis for this application in light of the nature of the work demands of such systems. Following that, a case study is presented that shows how this approach was applied to examine the training requirements of the F/A-18 system, a multirole fighter aircraft operated by the Australian Defence Organisation. The work domain analysis of this system is explained using the analytic themes in Chapter 8, so that another illustration of those guidelines is provided. Finally, this chapter discusses the usefulness and feasibility of work domain analysis for this application and notes some limitations of this case study.

Context

Establishing the training requirements of complex sociotechnical systems is essential for designing effective instructional systems. In this chapter, the application of work domain analysis to this problem is presented from the viewpoint of military procurement, which has the advantage of making this problem more tangible. Regardless, it is important to remember that the ideas put forward here are applicable to many kinds of systems, certainly not solely to military systems.

At the outset, it is worth clarifying that an effective instructional system is one that satisfies the training needs of the organization in question. This means that, to procure an effective instructional system, the requirements that it is developed to meet must be based on a systematic examination of the organization's training needs. Otherwise, there is a risk that the resulting instructional system, which may be acquired at considerable expense, will be unsuitable for the organization's training needs. In this chapter, the term *training needs* is used to refer to those work demands of a system for which workers require training. When developing an instructional system, therefore, a work analysis is essential for defining a system's work demands and, subsequently, examining its training needs and instructional system requirements. The term *training requirements* is used here to mean training needs and instructional system requirements collectively.

During the early stages of the military acquisition cycle, there are several occasions when the procuring organization may release instructional system requirements to the system manufacturers. The first time may be when the manufacturers are invited to declare their interest in developing or supplying the proposed instructional system. This 'invitation to register interest' is normally accompanied by a broad statement of requirement that provides a general idea of the scale and the nature of the instructional system sought by the procurer.

Having ascertained that there are one or more potential manufacturers for the instructional system, the procurer may issue a 'request for proposal' or 'request for tender.' The request for proposal, which calls on manufacturers to present an initial description of the instructional system they can offer, tends to be supplemented with a broad statement of requirement, similar to that provided with the original invitation. The request for tender, which calls on manufacturers to submit a comprehensive plan for the analysis, design, development, and delivery of the instructional system, is normally issued with a detailed statement of requirement.

After the procurer has selected a preferred manufacturer, a contract may be established between the two parties for the development of the instructional system in accordance with the statement of requirement. While it may be possible for the procurer to modify these requirements once the contract has been signed, or once the development of the instructional system has been initiated, these alterations may be costly to make. Generally, then, only minor refinements to the requirements are feasible at the later stages of the acquisition cycle. This makes it critical that the instructional system requirements correspond with the training needs of the relevant organization.

Rationale

Establishing the training requirements of complex sociotechnical systems presents distinct challenges for work analysis. One reason this is the case is that such systems are subject to disturbances, or unanticipated occurrences, that can affect their state (Vicente, 1999). As a result, while many of the events that workers in these systems have to deal with are familiar or well known to them, they also have to contend with unforeseen events, which pose significant threats to the systems' viability (Perrow, 1984; Rasmussen, 1969; Reason, 1990; Vicente, 1999). As explained in Chapter 1, workers often cannot rely on standard tasks or procedures for handling unanticipated events. Instead, flexible or innovative responses are usually required from them to manage these situations successfully. This means that the work demands of such systems cannot be characterized solely in terms of stable sets of tasks or task sequences for managing routine or predictable events (Meister, 1996; Rasmussen et al., 1994; Vicente, 1999). The training requirements derived from such descriptions of work demands would be incomplete.

Work domain analysis provides a framework for establishing the training requirements of complex sociotechnical systems that takes into account the nature of their work demands (Naikar and Sanderson, 1999). Specifically, this framework defines a system's work demands, not in the form of tasks or procedures, but in the form of the functional structure of the physical, social, or cultural environment of actors, which places constraints on their behavior. As these constraints are event independent, the training requirements derived from this form of analysis provide a basis for developing instructional systems that are suitable for training workers to deal with a variety of situations, including those that are unanticipated. Moreover, as these constraints can accommodate many different sequences of action, including those that cannot be specified a priori, the training requirements provide a basis for developing instructional systems that do not exclude flexible or innovative behavior but, in fact, support it explicitly. Work domain analysis, then, can be said to lead to the design of instructional systems with structural means–ends fidelity.

Figure 11.1 outlines how work domain analysis can be used to examine the training requirements of complex sociotechnical systems. First, a system's work demands may be specified in terms of the fundamental constraints on actors. The constraints in the figure are described in relation to the standard five levels of abstraction (Rasmussen, 1986; Rasmussen et al., 1994), which have been assigned the labels most recently put forward by Rasmussen (Reising, 2000).[*]

Once the system's work demands have been established, it becomes possible to consider whether workers require some form of training to satisfy them. In the first instance, each work demand, or constraint, only represents a potential training need, as workers may be capable of fulfilling some of the constraints without any training. Those work demands for which training is regarded as necessary signify the system's training needs. As shown in Figure 11.1, the constraints at each level of abstraction in a work domain model reveal qualitatively different training needs.

After the training needs have been ascertained, the instructional system requirements may be considered. As before, the constraints at each level of abstraction lead to qualitatively different requirements. Not all of the constraints may represent training needs and, therefore, instructional system requirements.

Case Study

This case study provides an account of how work domain analysis was applied to examine the training requirements of the F/A-18 system, a multirole fighter aircraft operated by the Australian Defence Organisation. The reason for this analysis was to facilitate the acquisition of a new instructional system. The aircraft, which is usually operated by a single pilot, has the capacity for launching attacks on both aerial and ground targets. Its suite of physical devices includes a multimode radar, global positioning and inertial navigation systems, a 'head-up' cockpit display, multifunction displays, and a range of weapons. The procurement of a new instructional system in this case was necessitated by a program of planned upgrades to the aircraft, which would significantly enhance its capability. The instructional system was expected to comprise a number of high-fidelity simulators.

Work Domain Analysis

In this section, the work domain analysis of F/A-18 is presented in the context of the analytic themes in Chapter 8. As was the situation with the preceding case studies, this analysis was conducted before the themes were devised. However, given that the analysis contributed to the development of the themes, presenting it in this form is straightforward. If readers are unconcerned with the details of how this analysis was carried out, they may move ahead to the next section, which explains how the resulting model was used for investigating the training requirements.

---

[*] See Chapter 3 for more detailed descriptions of the standard five levels of abstraction.

	Constraints	Potential Training Needs	Potential Instructional System Requirements
Functional Purposes	The objectives of a system and its external constraints	The objectives and external constraints that workers must be capable of fulfilling	The objectives and external constraints that an instructional system must have the capacity for training workers to fulfill
Value and Priority Measures	The criteria that must be respected for a system to achieve its functional purposes	The criteria that workers must be capable of respecting	The criteria that an instructional system must have the capacity for training workers to respect
Purpose-related functions	The functions a system must be capable of supporting to achieve its functional purposes	The functions that workers must be capable of managing	The functions that an instructional system must have the capacity for training workers to manage
Object-related Processes	The functional processes or functional capabilities or limitations of physical objects of relevance to a system	The functional processes or functional capabilities or limitations of physical objects that workers must be capable of exploiting	The functional processes or functional capabilities or limitations of physical objects that an instructional system must have the capacity for training workers to exploit
Physical Objects	The natural and artificial physical objects of relevance to a system	The natural and artificial physical objects that workers must be capable of recognizing	The natural and artificial physical objects that an instructional system must have the capacity for training workers to recognize

FIGURE 11.1
Application of work domain analysis for examining training requirements.

Theme 1: Purpose of the Analysis

The aim of the analysis was to assist the Australian Defence Organisation with examining the training requirements of F/A-18, particularly in connection with the pilots of the aircraft. The training requirements identified would inform the development of a broad statement of requirement for the new instructional system, which would be issued to industry manufacturers alongside the invitation to register interest. The scheme for how the work domain model would be used to establish the training requirements has already been outlined in general terms in this chapter.

Theme 2: Project Restrictions

The key project restrictions, which related to schedule, personnel, and finances, are described briefly here. How these restrictions influenced the way in which the analysis was performed and then applied is explained in subsequent themes.

The amount of time available for performing the analysis itself was approximately six months. In my role as a researcher at the Defence Science and Technology Organisation, I could allocate about half my time to this activity over that period. Prior to this project, I had some experience in conducting work domain analysis, specifically the original analysis of the Airborne Early Warning and Control (AEW&C) system reported in Chapter 9.

Approximately six months were available for using the resulting model to examine the training requirements and, subsequently, to contribute to the preparation of the statement of requirement for the instructional system. These activities were led by a subject matter expert, specifically an F/A-18 pilot based in the Capability Development Group, a component of the Australian Defence Organisation responsible for developing proposals and gaining government approval for future defense capabilities.

Theme 3: Boundaries of the Analysis

The focus system comprised the Australian Defence Organisation's F/A-18 system, which was the organizational entity demarcating the boundaries. The physical entity was a single aircraft, including its proposed upgrades, as well as its physical environment. The problem that was the focus of the analysis concerned the use of the aircraft to intercept or attack aerial and ground targets. This problem was modeled from the perspective of actors operating the aircraft.

Several factors were taken into account when selecting the physical entity that would delineate the focus system. Specifically, the boundaries included the proposed upgrades to the aircraft because the instructional system must have the capacity for training pilots to operate the modified or improved aircraft. The aircraft's physical environment was also incorporated into the model, given that the instructional system must have the capacity for training pilots to function within the constraints of this environment, such as those stemming from the terrain, weather, and friendly or hostile entities. The boundaries were limited to a single F/A-18 aircraft because other such aircraft that may be present within a formation can be represented as part of the physical environment in which the single aircraft operates. Taking this approach, the resulting training requirements would be the same for every pilot of any F/A-18 aircraft.

The focus problem reflected the fact that, instead of just flight training, the instructional system would be used to provide pilots with advanced tactical training. Therefore, rather than concentrating solely on the problem of flying the aircraft, the boundaries encompassed the broader problem of employing the aircraft to intercept or attack aerial and ground targets. It is also worth noting that, given that the instructional system was intended for training pilots in operating the aircraft, the focus system excluded problems not addressed within the aircraft, such as mission planning.

Theme 4: Multiple Models

Multiple models of stakeholders' perspectives could have been constructed for F/A-18, in particular, to depict the focus problem separately from the viewpoints of the pilots and the instructors. This approach would have recognized that the instructional system will be utilized not only by the pilots but also by the instructors. By capturing the perspectives of both those being trained and those providing the training, a comprehensive assessment of the training requirements could have been gained. As this approach was not practical in light of the project restrictions, however, only a single model was developed, which concentrated on the pilots' perspective of the focus problem. The work demands of the instructors were still considered in the development of the statement of requirement for the instructional system, although this was done by other means, including subject matter experts' judgments.

Multiple models of problem facets could also have been worthwhile in this case. The F/A-18 system may be viewed as loosely bound for reasons similar to AEW&C, as discussed in Chapter 9. Accordingly, by building separate representations of risks and resources, the focus problem could have been portrayed in depth as involving the distinct aspects of risk assessment and risk management. This approach could have enabled a thorough examination of the training requirements in relation to each facet, but, as it was not feasible within the project restrictions, only a single model was created. This model incorporated both risks and resources, so it still led to an examination of the training requirements in the context of both risk assessment and risk management, albeit in a less detailed and less explicit fashion. It also made the interactions between risks and resources more evident, which meant that it had the advantage of promoting an understanding of the training requirements arising from these interactions. The best solution might have been to adopt both approaches, but this was not viable within the project restrictions.

Theme 5: Causal–Intentional Continuum

The basis of the intentional and causal constraints pertinent to the focus system was found to be similar to that for AEW&C.[*] As a result, this focus system was also located toward the middle of the causal–intentional continuum[†] and classified as a loosely coupled intentional system, so that the model was defined principally by intentional constraints at higher levels of abstraction and causal constraints at lower levels.

Theme 6: Sources of Information

Documents were the main information sources for the analysis. Those particular to F/A-18 included flight manuals, tactical manuals, standard operating procedures, and student guides. More generally, the *Air Power Manual*, which describes the Royal Australian Air Force's doctrine, was also consulted for context. Knowledge-elicitation sessions with

---

[*] See Chapter 9 for a description of the intentional and causal constraints relevant to AEW&C.
[†] Figure 8.5 provides a graphical depiction of the causal–intentional continuum.

subject matter experts or field observations were not carried out specifically for this analysis. This decision was made in light of the project restrictions as well as the fact that I had previously held discussions with F/A-18 pilots and instructors and conducted observations of their training sessions.

Theme 7: Content of the Abstraction–Decomposition Space

The model of the focus system was defined by the standard five levels of abstraction (Rasmussen, 1986; Rasmussen et al., 1994).[*] During the project, these levels were assigned labels that were variations of those associated with the standard set. However, for the sake of consistency with the previous case studies, I refer to these levels with the latest labels proposed by Rasmussen (Reising, 2000).

The decomposition dimension was not specified in full because of the project restrictions. Instead, once the constraints in the model had been identified, only the physical objects and one of the purpose-related functions were represented at lower levels of detail using constraint decomposition.[†] The physical objects were depicted at several levels of decomposition simply because this was useful for handling the complexity of the physical components of the aircraft in building the model. The purpose-related function *communication and coordination* was decomposed to demonstrate the training requirements in relation to distributed mission training in greater detail.

Given that system decomposition had not been used to construct the model, the constraints were not identified by means of the full procedure discussed in Chapter 8 but on the basis of knowledge of the concepts that characterize the standard levels and structural means–ends relations. Specifically, these concepts and relationships provided a basis for analyzing material from documents, field observations, and subject matter experts for details about the constraints relevant to F/A-18. On the whole, information about the functional purposes was provided by accounts of why this system is present in the Australian Defence Force and how it contributes to Australia's defense capability. The value and priority measures could be extracted from explanations about the criteria used for judging the success of a mission and how the fulfillment of these criteria would be aided by the proposed upgrades to the aircraft. The purpose-related functions were apparent from descriptions of the functions or activities that the pilots perform on missions. Finally, information about the object-related processes and physical objects was readily available in accounts of the existing physical devices on the aircraft, the new physical devices that would be installed on the aircraft as a result of the planned upgrades, and the physical environment in which the aircraft operates.

Figure 11.2 portrays a sample of the constraints in the resulting model;[‡] the labels of some of the constraints have been simplified somewhat for ease of comprehension. A full glossary for the constraints in this model was not developed because of the project restrictions. However, many of the constraints were described in more depth by collating relevant excerpts about the system from the various information sources.

---

[*] As noted in Chapter 9, I would now tend to define the abstraction dimension without presuming the standard levels, by following the guidelines in Chapter 8.

[†] As stated in Chapter 9, I would now use system decomposition to specify the decomposition dimension, so that a skeletal abstraction–decomposition space could be assembled, the benefits of which are described in Chapter 4.

[‡] I would now represent constraints with labels that emphasize the functional structure of the environment, as indicated in Chapter 9.

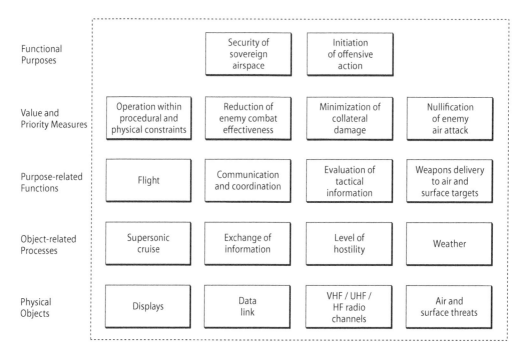

FIGURE 11.2

Sample of constraints in the F/A-18 model. HF, high frequency; UHF, ultrahigh frequency; VHF, very high frequency.

Theme 8: Validity of the Model

The comprehensiveness and accuracy of the model were established by having it reviewed by subject matter experts. The review entailed asking experts a number of questions about the content of the model in the form of semistructured interviews. Because of the project restrictions, only one F/A-18 training instructor and three air combat operations analysts participated in the validation exercise. Although all of these experts were shown the entire model during the review, the elements they were asked to comment on depended on their knowledge and experience. The experts verified many aspects of the model but also pointed out some minor problems with it, such as inaccurate or missing information. Accordingly, the model was refined following the review.

In its completed form, the model contained 5 functional purposes, 11 value and priority measures, 12 purpose-related functions, 20 object-related processes, and 63 physical objects. These constraints were linked by approximately 160 structural means–ends relations. The links encompassed such instantiations as those that were already engineered into the system, those that would be built into the system as a result of the proposed upgrades, those that subject matter experts either knew about or could foresee, and those that were evident in light of particular situations.[*]

Training

Having considered how the work domain model of F/A-18 was produced, it is now possible to explain how this model was used for examining the training requirements of

---

[*] Chapter 5 discusses the concept of instantiations of structural means–ends relations in greater depth.

the pilots of this aircraft. Before doing this, it should be noted that aside from the work domain analysis, a training task analysis was performed on this project with the same intention of identifying the training requirements of F/A-18; the reasons that both analyses were carried out are addressed later in the chapter. The training task analysis, which was conducted by Wallace, Walkden, and Lintern (1999), involved the development of a comprehensive taxonomy of tasks in the form of a three-tiered hierarchy. At the first level, the system's major processes (i.e., normal operations, tactical operations) were specified. The second level described segments (e.g., circuits, intercepts) within each of the major processes. Then, at the last level, events within each of the segments requiring a skilled response from pilots (e.g., engine failure after take-off, stern conversion intercept) were identified. The final taxonomy consisted of 1,136 tasks, which were used as a basis for defining the training requirements.

The process for using the work domain model to examine the training requirements involved two main stages. First, the constraints in the model were reviewed to establish which of them represent actual, rather than potential, training needs; as discussed previously, not all of the constraints necessarily reflect work demands for which workers require training. Input from subject matter experts, specifically F/A-18 pilots and instructors, was necessary for this stage.

Subsequently, the instructional system requirements for each of the training needs were explored. This involved, in essence, investigating what capabilities the instructional system must have in order to fulfill the training needs. In this case, three main types of requirements were considered, namely, capabilities for simulation, scenario generation, and trainee assessment. This stage was also achieved in consultation with subject matter experts, who included not only F/A-18 pilots and instructors but also scientists and engineers working in the areas of training and simulation.

Figure 11.3 illustrates how the F/A-18 model can be used as a basis for examining the training requirements. The constraints listed in this figure against each level of abstraction are the same as those shown in Figure 11.2. These constraints may be viewed in terms of potential training needs and instructional system requirements, which are qualitatively different at each level of abstraction.

As a case in point, the constraints at the physical objects level signify that pilots must be capable of recognizing the functionally relevant properties of natural and artificial entities in their environment. For example, they must be familiar with the layout of various *displays* in the aircraft's cockpit as well as visual indicators that characterize different aerial *threats*. These work demands may be found to represent actual training needs and instructional system requirements. Accordingly, the capabilities required for simulation, scenario generation, and trainee assessment may be defined. For instance, the instructional system must have the capacity to simulate or re-create the arrangement of displays in the aircraft's cockpit and the visual markings of aerial threats. In addition, it must support the development of scenarios that are suitable for training pilots to meet these visual work demands and enable their performance in relation to these work demands to be evaluated.

Similarly, at the value and priority measures level, the constraints signify that pilots must be capable of respecting a range of criteria to fulfill the system's functional purposes. Consider, for instance, the criteria of *reduction of enemy combat effectiveness* and *minimization of collateral damage*. Alongside the purpose-related function of *weapons delivery to air and surface targets*, these criteria suggest that if pilots cannot limit the enemy's combat potential without causing unnecessary or excessive damage when delivering weapons to a designated target, the functional purposes will not be achieved. Given that these work demands denote actual training requirements, it may be ascertained

	F/A-18 Constraints	F/A-18 Potential Training Needs	F/A-18 Potential Instructional System Requirements
Functional Purposes	Security of sovereign airspace, Initiation of offensive action	Pilots must be capable of fulfilling objectives and external constraints such as security of sovereign airspace and initiation of offensive action	The instructional system must have the capacity for training pilots to fulfill objectives and external constraints such as security of sovereign airspace and initiation of offensive action
Value and Priority Measures	Operation within procedural and physical constraints, Reduction of enemy combat effectiveness, Minimization of collateral damage, Nullification of enemy air attack	Pilots must be capable of respecting criteria such as operation within procedural and physical constraints, reduction of enemy combat effectiveness, minimization of collateral damage, and nullification of enemy air attack	The instructional system must have the capacity for training pilots to respect criteria such as operation within procedural and physical constraints, reduction of enemy combat effectiveness, minimization of collateral damage, and nullification of enemy air attack
Purpose-related Functions	Flight, Communication and coordination, Evaluation of tactical information, Weapons delivery to air and surface targets	Pilots must be capable of managing functions such as flight, communication and coordination, evaluation of tactical information, and weapons delivery to air and surface targets	The instructional system must have the capacity for training pilots to manage functions such as flight, communication and coordination, evaluation of tactical information, and weapons delivery to air and surface targets
Object-related Processes	Supersonic cruise, Exchange of information, Level of hostility, Weather	Pilots must be capable of exploiting functional processes or functional capabilities or limitations of physical objects such as supersonic cruise and exchange of information. In addition, pilots must be capable of exploiting conditions created by physical objects such as level of hostility and weather	The instructional system must have the capacity for training pilots to exploit functional processes or functional capabilities or limitations of physical objects such as supersonic cruise and exchange of information. In addition, the instructional system must have the capacity for training pilots to exploit conditions created by physical objects such as level of hostility and weather
Physical Objects	Displays, Data link, VHF/UHF/HF radio channels, Air and surface threats	Pilots must be capable of recognizing the functionally relevant properties of natural and artificial physical objects such as those relating to displays, data link, VHF/UHF/HF radio channels, and air and surface threats	The instructional system must have the capacity for training pilots to recognize the functionally relevant properties of natural and artificial physical objects such as those relating to displays, data link, VHF/UHF/HF radio channels, and air and surface threats

FIGURE 11.3
Sample of potential training requirements for F/A-18.

that the instructional system must have the capacity to simulate the extent of damage that pilots cause when delivering weapons to a target. It must also allow scenarios to be developed that provide pilots with a range of opportunities, at varying levels of difficulty, to practice limiting collateral damage when delivering weapons to a target, while still degrading the enemy's combat potential. Furthermore, the instructional system must have the capacity to enable the assessment of pilots' performance in meeting all of these work demands.

It is worth pointing out that a work domain model can be used to derive training requirements at a more detailed level than that outlined here. In the case of F/A-18, however, the objective of the model was to inform the development of a broad statement of requirement for the new instructional system, which would accompany the invitation to register interest issued to industry manufacturers. The training requirements highlighted by this model were sufficiently detailed for this purpose. Also, it was not practical to develop training requirements in more depth given the project restrictions.

If the analysis had been undertaken during the later stages of the acquisition cycle, it might have been necessary to produce more detailed training requirements, for example, to supplement the request for proposal or request for tender. This could have been achieved by modeling F/A-18 at a finer level of granularity. This step, however, was not required since the Australian Defence Organisation opted to commit to a particular manufacturer's instructional system following the invitation to register interest. Even so, it should be acknowledged that a more fine-grained model would not have specified the training requirements completely. For instance, the levels of fidelity necessary to simulate the functionally relevant properties of physical objects could not have been identified. Input from subject matter experts, other forms of work or task analysis, or experimental studies are required to achieve this level of specification.

Impact

Whether work domain analysis provides a useful approach for examining training requirements can be judged on the basis of its impact or ability to influence practice. On the F/A-18 project, the impact of this approach was evident in two significant ways.

First, a decision was made by the Capability Development Group and the F/A-18 System Program Office, the two main units concerned with the procurement of the new instructional system, to adopt work domain analysis as an additional approach for examining training requirements. Originally, only the training task analysis was to be employed for this purpose. The main rationale for the decision to adopt both approaches was that, together, they would offer a more comprehensive strategy for defining the statement of requirement for the instructional system than that offered by the training task analysis alone. Whereas the training task analysis would identify the training requirements in relation to a set of tasks or procedures for dealing with routine or familiar situations, the work domain analysis would highlight the training requirements in light of a range of constraints, which are event independent. Consequently, the level of risk in the acquisition of the instructional system would be reduced.

Further evidence of impact was that the training requirements revealed by the work domain model were used, along with those from the training task analysis, to contribute to the preparation of a broad statement of requirement for the instructional system. This statement of requirement was released together with the invitation to register interest that was issued to industry manufacturers.

Uniqueness

Another basis for judging the usefulness of work domain analysis for examining training requirements is with regard to its uniqueness relative to standard techniques. A common technique used in industry is the instructional systems development process (Childs and Bell, 2002; Department of the Air Force, 1993; Dick and Carey, 1990; Dick, Carey, and Carey, 2005; Gagne, Wager, Golas, and Keller, 2005; Hays and Singer, 1989). This technique may be characterized as involving four main phases: analysis, design, development, and delivery (Childs and Bell, 2002). A critical component of the first phase is a task analysis, the objective of which is to produce descriptions of a system's work demands that can be used as a basis for ascertaining its training requirements (Hays and Singer, 1989).

Hays and Singer (1989) identify six categories of task analysis techniques that have been used for developing instructional systems. These are the behavior-description approach, the behavior-requirements approach, the abilities-requirements approach, the task-characteristics approach, the phenomenological approach, and the information-theoretic approach. Although these approaches have slightly different foci or emphases, as indicated by their labels, they all involve the analysis of tasks or task sequences.

Standard techniques, then, base the training requirements of complex sociotechnical systems on specifications of their work demands in the form of a set of tasks or procedures. However, as pointed out previously, such descriptions can only be produced in relation to events that are known or that can be anticipated.* While a large proportion of the events faced by workers do fall into this category, the very nature of complex sociotechnical systems means that workers must also deal with novel events, which are often hazardous. As a rule, novel events cannot be handled effectively with routine or familiar procedures but, instead, require creative responses from workers, which cannot be specified in advance. This means that specifications of work demands solely in the form of stable tasks or task sequences are bound to lead to training requirements that are incomplete.

In the case of F/A-18, work domain analysis complemented the training task analysis that was also performed on this project and made a unique contribution to the examination of training requirements. The work domain model identified the work demands of this system in the form of a set of constraints, which are event independent. As a result, the training requirements revealed in this context are pertinent to many kinds of situations, including unforeseen events. Furthermore, as the constraints encompass multiple possibilities for action, the training requirements support creative behavior. In this way, work domain analysis promotes the design of instructional systems with structural means–ends fidelity.

It is important to point out that standard techniques and work domain analysis may sometimes appear to be concerned with the same information for establishing training requirements. At one level, this conclusion is true. For example, like work domain analysis, standard techniques may be concerned with the functionality of physical devices or the criteria for evaluating trainees' performance. However, as noted before, there is a fundamental difference between the two approaches. Continuing with the preceding examples, standard techniques focus on examining the functionality of physical devices that is required for performing particular tasks or procedures, which can only be described for known or anticipated events. In contrast, work domain analysis concentrates on investigating the full range of functionality that is afforded by the physical devices, which remains

---

* Chapters 1 and 7 explain why tasks or task sequences can only be specified in relation to known or anticipated events.

constant irrespective of the situation. While some of the functionality uncovered by this form of analysis may not be necessary for handling routine or familiar events, it may be critical for dealing with novel situations. The training requirements revealed by work domain analysis account for this possibility. Similarly, whereas standard techniques concentrate on identifying criteria for evaluating trainee performance on specific tasks or procedures, work domain analysis focuses on defining the criteria that must be respected by workers regardless of the situation.

Finally, it may be argued that techniques such as hierarchical task analysis, cognitive task analysis, or functional decomposition could be used to uncover the same set of training requirements as work domain analysis. However, unlike work domain analysis, none of the other techniques have a conceptual basis that is unequivocally concerned with defining a system's work demands in a way that is relevant to a broad range of situations, including those that are unpredictable. Therefore, unless the other techniques are somehow inherently capable of meeting this requirement, despite not having this conceptual focus, or their application is informed by the principles of work domain analysis, it seems unlikely that they would arrive at the same set of training requirements.

Feasibility

The feasibility of work domain analysis for examining training requirements may be established with respect to whether it can be achieved within a project's schedule, personnel, and financial restrictions. This was certainly found to be the case for F/A-18. The schedule and personnel limits were noted previously. As observed with regard to the preceding case studies, the main financial cost stemmed from the wages of the analyst and the subject matter experts, but this outlay was insignificant compared with the project's overall budget.

Limitations

The limitations of this case study are similar to those of the previous case studies in that it does not provide empirical evidence that work domain analysis leads to the development of effective instructional systems or to better instructional systems compared with those resulting from standard techniques. The sort of study required to obtain such evidence is impractical in industrial settings.

To establish whether work domain analysis resulted in an effective instructional system for F/A-18, it would have been necessary to develop a solution based solely on the training requirements identified by this technique. Following that, the performance of pilots trained with this solution would have had to be evaluated. Similarly, to determine whether work domain analysis led to a better instructional system than that resulting from the training task analysis that was also performed on this project, it would have been necessary to develop two solutions, one based exclusively on each technique. Subsequently, the performance of pilots trained with each solution would have needed to be compared. Such studies were not feasible on this project.

As pointed out in relation to the previous case studies, these kinds of impediments to the empirical validation of techniques in industrial settings are not specific to work domain analysis. This explains why techniques implemented in industry are usually not evaluated in a formal sense (Czaja, 1997) but, instead, tend to be assessed against the criterion of usefulness (Vicente, 1999; Whitefield et al., 1991). In the case of F/A-18, the usefulness of work domain analysis for examining training requirements was apparent in terms of its impact and uniqueness, and its feasibility was confirmed as well. To establish whether this approach is useful and feasible consistently, future research should focus on adding to this body of evidence by contributing additional case studies.

Summary

This chapter focused on the problem of establishing the training requirements of complex sociotechnical systems. This is an important consideration for developing effective instructional systems, that is, systems that are suitable for satisfying the training needs of the organizations for which they have been created. Work domain analysis offers a valuable approach for addressing this problem because, by defining training requirements on the basis of a set of constraints, it promotes the design of instructional systems with structural means–ends fidelity. These are systems that can be used to train workers to deal with a wide range of situations, both anticipated and unanticipated, specifically by providing them with opportunities to practice responding to these situations in flexible and innovative ways. A case study was provided that showed how work domain analysis was used to examine the training requirements of F/A-18, a multirole fighter aircraft operated by the Australian Defence Organisation. Aside from illustrating the guidelines for work domain analysis in Chapter 8, this case study demonstrated the usefulness and feasibility of this approach for yet another novel application, which confirms its value for applications beyond designing ecological interfaces.

Conclusion

This note draws to a close the four main parts of this book. In the first part, readers were introduced to cognitive work analysis and work domain analysis. The second part explained the concepts of work domain analysis in more depth, including those that had not been addressed in detail—or at all—by the earlier foundational texts. The third part provided guidelines for performing this kind of analysis, in the form of a set of analytic themes, which placed the material from the earlier parts into a methodological context. In the fourth part, three case studies of the application of this approach to large-scale military systems were presented. As well as providing comprehensive illustrations of the application of the guidelines, these case studies emphasized the significance of this form of analysis for attending to a variety of problems in industrial settings besides interface design.

One more part of this book, the Appendix, remains. This part serves the purpose of reminding readers that work domain analysis is only one dimension of cognitive work analysis. To this end, the key concepts of the remaining dimensions—control task analysis, strategies analysis, social organization and cooperation analysis, and worker competencies analysis—are described briefly.

Overall, the objective of this book is to make work domain analysis more accessible to readers. This objective was not reached by oversimplifying the principles of this form of analysis. Instead, as will now be evident, its conceptual foundations and methodology were treated in great depth, and this material was illustrated with numerous examples spanning a range of systems, including one that will be highly familiar to readers, namely, a home. As stated at the outset, I believe a sound appreciation of work domain analysis will lead to more powerful and innovative applications of this approach as well as make it easier and more enjoyable to perform.

That said, work domain analysis is an inherently challenging undertaking, particularly when applied to large-scale systems. This technique is founded on complex concepts that require hard work and dedication, not only to understand but also to put into practice effectively. The challenging nature of the exercise also arises from the complexity of the systems we analyze. Given the goal of designing successful work practices, and creating systems that are safer and healthier for workers, this effort is worthwhile.

Section V

Appendix

The Remaining Dimensions of Cognitive Work Analysis

OVERVIEW Having dealt with work domain analysis comprehensively in the body of this book, in this appendix I provide a brief description of the key concepts of the other dimensions of cognitive work analysis, namely, control task analysis, strategies analysis, social organization and cooperation analysis, and worker competencies analysis.[*] These concepts, which were derived predominantly from the texts by Rasmussen et al. (1994) and Vicente (1999), are also illustrated with examples from a home. This appendix makes clear that not all of the requirements for designing effective systems are addressed with work domain analysis. Some of these requirements are considered within the other dimensions.

Control Task Analysis

Control task analysis is concerned with the constraints that are placed on actors by *what* needs to be done in a system. Specifically, these constraints arise from the activity that is necessary to achieve the system's purposes with a given set of physical resources. The modeling tools that are relevant for this dimension are the contextual activity template[†] and the decision ladder template.[‡]

Unlike work domain analysis, control task analysis is event *dependent* since it focuses on defining what needs to be done in known, recurring classes of situations. This fundamental difference between the two dimensions may be illustrated by revisiting the analogy of an ant on the beach.[§] Work domain analysis, in the context of this analogy, focuses on the constraints that the beach places on the ant. These constraints shape the ant's behavior irrespective of the situation. In contrast, control task analysis is concerned with the constraints that shape the ant's behavior when it is pursuing certain goals on the beach, such as when it is going home or when it is foraging for food (Vicente, 1999). These constraints are dependent on the situation. Whereas the objective of work domain analysis is to support workers in handling a broad range of situations, including unanticipated events, the aim of control task analysis is to support workers in dealing with known, recurring classes of situations.

While control task analysis focuses on what needs to be done, it is not concerned with how or by whom this activity can be done. These other considerations are treated independently by strategies analysis and social organization and cooperation analysis, respectively. Postponing decisions relating to these matters until the later dimensions of the framework offers leverage points for design (Naikar, 2006a, 2010; Vicente, 1999).

---

[*] This appendix is based on a paper published by Naikar (2006b).
[†] See Figure 1.5a for the contextual activity template.
[‡] See Figure 1.5b for the decision ladder template.
[§] Chapter 2 uses the analogy of the ant on the beach to describe work domain analysis.

To explain, cognitive work analysis takes account of the fact that, typically, a range of strategies and organizational structures is relevant for performing a single activity. Furthermore, workers tend to switch between the different strategies and organizational structures to suit the situational demands. For example, if a situation changes suddenly from routine to atypical, workers may switch from using mental simulation for evaluating a single course of action to adopting an analytical strategy for comparing several options. By first establishing solely what needs to be done in a system, it becomes possible, later in the analysis, to consider a range of alternatives for both how and by whom the activities can be performed. Consequently, designs may be developed that support workers in shifting flexibly between multiple strategies or organizational structures in response to the situational demands.

Cognitive work analysis also recognizes that in complex sociotechnical systems, the same goals may be accomplished with numerous, perhaps infinite, task or action sequences, and that the sequences that are executed may vary as a function of several factors, including the preferences of individual workers. Therefore, for modeling activity in these systems, a form of analysis that identifies what needs to be done without specifying particular sequences of tasks or actions is appropriate. For this reason, control task analysis does not decompose activity into such idiosyncratic sequences but, instead, characterizes it in the form of a set of recurring work situations, work functions, or control tasks (Naikar et al., 2006; Rasmussen et al., 1994; Vicente, 1999).

In its first stage, control task analysis models what needs to be done in a system in terms of a set of work situations or work functions; Rasmussen et al. (1994) call this stage *activity analysis in work domain terms*. Work situations are relevant when activity in the system is structured on the basis of time or space. In contrast, work functions are appropriate when activity is organized in line with its functional content, independently of its temporal or spatial characteristics. In some systems, activity is best characterized as a combination of work situations and work functions (Naikar et al., 2006). These are systems in which activity within a work situation, or set of work situations, is further ordered according to its functional content. In this case, the combinations of work situations and work functions that are possible may be represented with a contextual activity template.

Figure A.1 presents a hypothetical contextual activity template of a home. The work situations, which are different stages of the day, are shown along the horizontal axis. The work functions, which are different types of problems, are portrayed along the vertical axis. Within the template, the boxes indicate the work situations in which a work function *can* occur, whereas the bars depict the work situations in which a work function *typically* occurs. The combinations of work situations and work functions that are possible impose qualitatively different cognitive demands on actors.

In the next stage of control task analysis, activity is characterized in terms of the control tasks that are required for each work situation or work function. Rasmussen et al. (1994) call this stage *activity analysis in decision making terms*. For modeling control tasks, the decision ladder template is useful.

Figure A.2 presents a hypothetical decision ladder for a home, specifically for the work function of *prepare meals and beverages*. The annotations to this decision ladder are in the form of questions, which reflect the recurring concerns of actors (Elix and Naikar, 2008a). The answers to these questions, which are formulated by actors in the context of particular circumstances, reflect their states of knowledge. Different control tasks, or parts of the decision ladder, may be active in different work situations. Thus, overall, many combinations of work situations, work functions, and control tasks are possible, which place qualitatively distinct cognitive demands on actors.

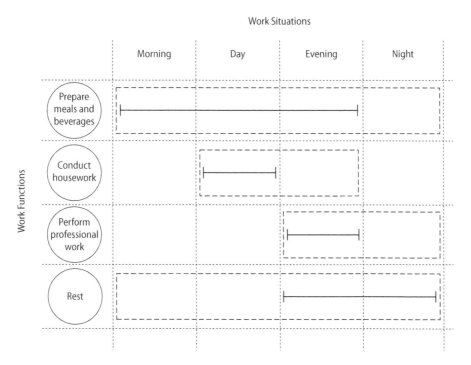

FIGURE A.1
Hypothetical contextual activity template of a home.

As shown in Figure A.2, the decision ladder template is composed of links between boxes and ovals. The boxes denote information-processing activities, whereas the ovals signify states of knowledge that are the inputs or the outputs of those activities. The left arm of the decision ladder comprises control tasks relating to the observation of information and situation analysis, whereas the right arm consists of control tasks connected with planning and execution. Last, the top part of the decision ladder is defined by control tasks involving option evaluation and goal selection.

The decision ladder template may appear to resemble traditional models of human information processing. However, there are three main differences between such models and the decision ladder. First, the decision ladder need not be followed in a linear sequence. Instead, shortcuts are possible from one part of the template to another. (Some examples of shortcuts are indicated by the dashed arrows in Figure A.2.) Second, activity need not start in the bottom left box and finish in the bottom right box; rather, different start and end points are possible. Third, activity may flow from right to left, not just from left to right. Hence, on the whole, the decision ladder reflects Rasmussen's (1974) observations that experts can string together control tasks in many different ways, depending on the demands of the situation.

To conclude, control task analysis recognizes that expertise is a constructive process that involves generating *"a contextually tailored sequence of cognitive activities* that is appropriate for the present situation" (Vicente, 1999, p. 186; italics in the original). As suggested above, in many systems numerous combinations of work situations, work functions, and control tasks are possible. The particular combination that actors adopt will vary as a function of the circumstances. Designs, therefore, must provide actors with the flexibility to tailor their activities to local contingencies.

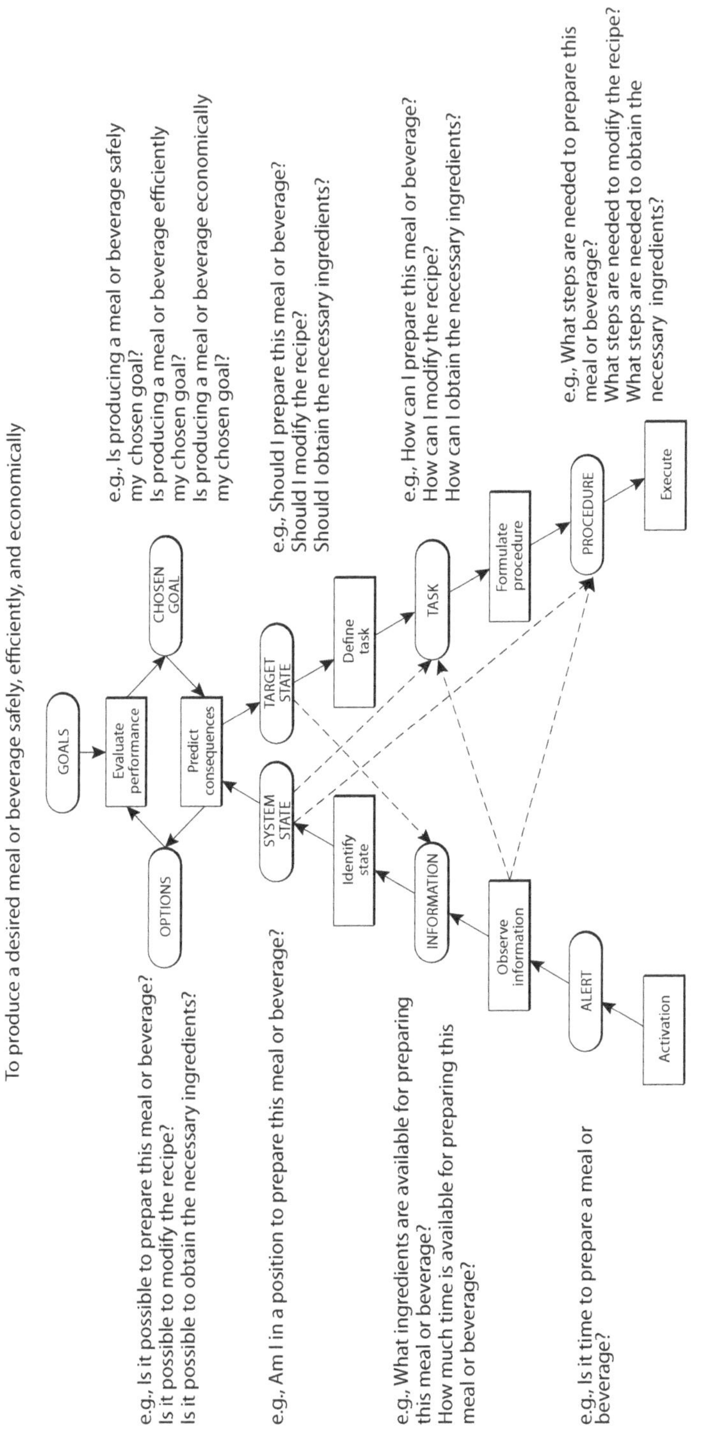

FIGURE A.2
Hypothetical decision ladder for the work function of *prepare meals and beverages*.

Strategies Analysis

Strategies analysis focuses on the constraints that are imposed on actors by the different ways in which activity can be accomplished in a system. Thus, whereas control task analysis is concerned with *what* activity is needed, strategies analysis is concerned with *how* the activity can be done. An information flow map may be used to produce a graphical representation of a strategy.[*] However, as yet, this tool has not been developed in a generic format but has been created only for specific applications (Vicente, 1999).

In the context of this form of analysis, strategies are not viewed as detailed sequences of actions or mental operations. Instead, they are envisaged as categories of cognitive procedures, which are idealized, abstract descriptions of particular processes.

Given that several strategies are usually possible for accomplishing a single activity, this dimension involves examining the range of strategies that are applicable for particular activities. For instance, a number of strategies may be used to prepare a meal in a home, including (1) following a written recipe; (2) preparing a familiar meal from memory; (3) creating a meal on the basis of one's knowledge of the functional properties of ingredients, such as their tastes, aromas, textures, and cooking times; (4) creating a meal by trial and error; and (5) preparing a meal by heating up a precooked or frozen dish.

Actors, then, can choose between multiple strategies for carrying out an activity. The strategies they select at any point in time will depend on the *task demands* and the *performance criteria*, or resource requirements, of each strategy, such as the amount of time, memory, or knowledge that is required (Rasmussen et al., 1994). Preparing a meal by heating up a precooked dish requires less time than does creating a meal on the basis of one's knowledge of the functional properties of ingredients, so the former strategy is more likely to be adopted when time is limited. *Actors' subjective task formulation*, or how actors conceive of a task, may also affect which strategy is chosen (Rasmussen et al., 1994). A person who formulates the act of preparing a meal as a chore that is necessary to satisfy hunger is likely to adopt a different strategy from someone who views it as a creative process that is challenging and rewarding.

The preceding examples do not illustrate that actors will often switch between multiple strategies while performing an activity to suit the situational demands. For instance, actors may begin to prepare a familiar meal from memory but then refer to a cookbook to check aspects of the recipe that they have forgotten. Alternatively, they may start with a written recipe to prepare a meal but then add some extra ingredients, which would otherwise be wasted, to the recipe on the basis of their knowledge of the functional properties of the ingredients.

In the case of any system, it is difficult to anticipate when strategy shifts will occur and which strategies will be selected at any moment. However, the goal of strategies analysis is not to make these kinds of predictions. Instead, by identifying the design requirements of multiple strategies, the goal is to design systems that will support workers in performing their jobs "in a flexible manner by using whatever strategy they prefer, and by seamlessly switching between strategies as necessary" (Vicente, 1999, p. 222).

Finally, it is important to emphasize that this dimension is concerned with uncovering the strategies that are possible independently of who will execute them. It is particularly critical to avoid the mistake of simply studying the set of strategies that workers use. Workers might disregard strategies that are resource intensive but, as a result, neglect some very effective practices. The resource requirements of a strategy can be manipulated

---

[*] Figure 1.6 presents an information flow map.

through design, for example, by creating effective displays or by offloading the demanding aspects of a strategy to automation. Consequently, workers will be able to adopt strategies they otherwise might not use.

Social Organization and Cooperation Analysis

Social organization and cooperation analysis focuses on the constraints that are placed on actors by the allocation, distribution, and coordination of work in a system. From the standpoint of this form of analysis, a system's work demands can be allocated to individual workers, to teams of workers, or to automation. The modeling tools from the preceding dimensions are useful for this dimension as well. Rasmussen et al. (1994) and Vicente (1999) describe this form of analysis in a manner that emphasizes its suitability for defining, and thus supporting, existing organizational structures. Here, this dimension is portrayed in a way that highlights its relevance for designing new organizational structures.

Underpinning this dimension is the view that in complex sociotechnical systems, flexible organizational structures that can be adapted to local contingencies, including unforeseen events, are essential for effective performance. Consequently, there is no attempt in this form of analysis to prescribe a single, or best, organizational structure. Emphasis is placed, instead, on defining the set of possibilities for work organization in a system and developing a team or organizational design that can support or adopt those possibilities.

To examine the possibilities for work organization in a system, it is necessary to identify the *criteria* that may shape, or govern, how work is allocated or distributed across actors. Rasmussen et al. (1994) and Vicente (1999) describe six potential criteria. These criteria, which are noted here with examples from a home, are (1) actor competency—the competencies of actors may shape how responsibilities for preparing meals versus gardening are distributed; (2) access to information or means for action—the access that actors have to information, such as the ingredients required for a meal, or to the means for action, such as a car, may influence who is given the responsibility for grocery shopping; (3) coordination—the requirement to minimize communication may shape the allocation of housework to a single actor; (4) workload—the need for workload sharing may influence the distribution of housework across multiple actors; (5) safety and reliability—the requirements for safety and reliability may shape the responsibilities given to children; and (6) compliance—the need for compliance with established practices or regulations may shape the hiring of qualified technicians for electrical repairs.

Once the criteria have been identified, it is necessary to examine how the system's work demands may be distributed across actors given those criteria. The criteria may be applied singly or in various combinations. The distribution of work across actors, as a function of the criteria, may be considered in relation to the work domain, work situations, work functions, control tasks, or strategies. That is why the modeling tools from the preceding dimensions are also relevant here.

Figure A.3 illustrates how the decision ladder for the work function of *prepare meals and beverages* may be used to examine the distribution of control tasks across actors. Given certain criteria, such as the necessity for workload sharing or the competencies of actors, the shaded areas of the model might be allocated to Actor A and the remaining areas to Actor B. As a result, the nature of the work content, or the responsibilities, of the two actors would be revealed. Whereas Actor A has responsibility for deciding which meals or

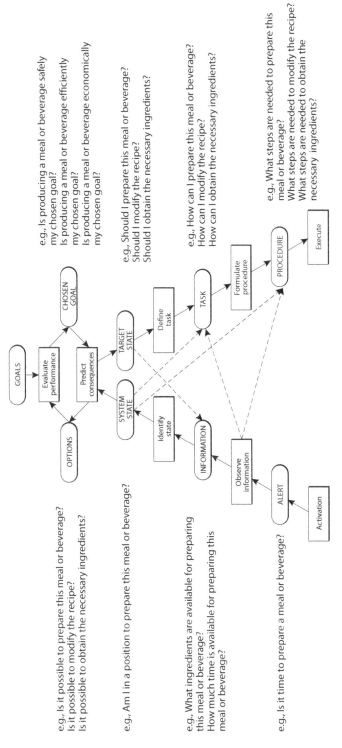

FIGURE A.3
Use of the decision ladder to examine possibilities for work organization.

beverages to prepare, Actor B is responsible for preparing the selected items. This representation also reveals the information, coordination, and resource requirements of the two actors, given their work content. For instance, to prepare a meal or beverage, Actor A must have access to information about the available ingredients, Actor A must communicate this information to Actor B, and Actor B must have access to the necessary ingredients. The other modeling tools of cognitive work analysis may be used in a similar fashion to identify the boundaries of responsibilities of actors, and their total set of information, coordination, and resource requirements, given the criteria for work allocation.

Once the set of possibilities for work organization in a system has been defined, the next step is to identify the team or organizational design requirements for each possibility. For example, the requirements may relate to the team size, hierarchical structure, subgroups, or roles necessary for sustaining each possibility. Following that, a team or organizational design that best fulfills the set of requirements, or that best supports the range of possibilities for work organization, may be created.

Aside from the possibilities for work organization, this dimension is concerned with the form of communication, or social organization, adopted for coordination in a system. Some examples of forms of social organization are autocratic, authoritarian, and heterarchical (Rasmussen et al., 1994; Vicente, 1999). In an autocratic organization, a single actor takes charge of all of the other actors, whereas an authoritarian organization is characterized by clearly defined chains of command. In contrast, a heterarchical organization is one that can take many forms, including democratic, whereby decision making occurs by committee, and anarchic, whereby individuals act independently, voluntarily, and cooperatively. The form of social organization in a system may be adapted to local contingencies. For instance, a democratic organization might adopt an autocratic form in the event of an emergency.

It is important not to be misled into supposing that because this dimension is concerned with examining the possibilities for work and social organization in a system, its aim is to plan a priori the nature of the organizational structures that should be adopted by actors in different situations. On the contrary, this notion is avoided deliberately because it is clear that in highly effective and reliable systems, the organizational structures that are brought into play are generated online and in real time by multiple, cooperating actors responding to the local context. The aim, therefore, of identifying the total set of possibilities for work allocation, distribution, and coordination is to promote flexibility and adaptation in organizations by developing designs that are tailored to the requirements of the various possibilities.

Worker Competencies Analysis

Worker competencies analysis is concerned with the constraints that are placed on actors by the competencies required for dealing with a system's work demands. These work demands are the subject of the preceding dimensions. Human cognitive capabilities and limitations place constraints on how these work demands can be met. For this dimension, the skills, rules, and knowledge taxonomy[*] provides a tool for integrating information about a system's work demands with knowledge about human cognition in a manner that brings to light the implications for design.

---

[*] The skills, rules, and knowledge taxonomy is depicted in Figure 1.8.

Key to this form of analysis is the recognition that the work demands of a system *collectively* shape the competencies that workers need. For example, if certain control tasks (control task analysis) are always allocated to automated devices (social organization and cooperation analysis), workers will require a different set of competencies from those necessary if they are to perform those control tasks themselves. The first step for this dimension, then, involves consolidating the work demands from the preceding dimensions.

Following that, the skills, rules, and knowledge taxonomy is used to consider three categories of human behavior, or three levels of cognitive control, that workers may adopt for fulfilling the system's work demands. Skill-based behavior is defined by highly automated and integrated patterns of action that are performed in real time and are directly coupled to the environment in a continuous perception–action loop. This form of behavior is executed without conscious attention and, as a result, cannot be verbalized. Chopping vegetables with a knife, for example, involves skill-based behavior.

Rule-based behavior is characterized by if–then mappings between familiar perceptual cues in the environment and appropriate actions. The rules for behavior can be derived from procedures, experience, instruction, or prior problem-solving activities (Vicente, 1999). Reasoning is not required in this form of behavior since cues in the environment trigger actions directly and goals are not considered explicitly. However, workers are usually conscious of their cognitive activities and, therefore, can verbalize their thoughts. Following directions in a recipe to switch the stove to a low heat setting when the contents of a pot start to boil involves rule-based behavior.

Knowledge-based behavior is defined by serial, analytical reasoning based on a symbolic mental representation of the relevant goals and constraints in the environment. This form of behavior requires conscious, focal attention and is slow, effortful, and deliberate. It is usually associated with unfamiliar situations, when prior experience is inapplicable. Substituting an ingredient in a recipe with another ingredient by reasoning about their functional properties (e.g., tastes, aromas, textures, cooking times) involves knowledge-based behavior.

All three levels of cognitive control can be active at any moment. For instance, people may chop vegetables with a knife while thinking about what substitute they might use for a particular ingredient in a recipe and waiting for the contents of a pot to start to boil. Hence, there are significant interactions between the three levels.

The level of cognitive control that is activated in performing a task depends on how information in the environment is *interpreted* by workers. Information can be interpreted in terms of signals, signs, or symbols. Signals, which are continuous and quantitative perceptual indicators of time–space patterns, trigger skill-based behavior. For example, when pouring milk into a measuring jug to obtain the amount of a cup, the decreasing distance between the level of liquid in the jug and the one-cup marker is a time–space signal that indicates when to stop pouring. Signs, which are arbitrary but familiar perceptual cues in the environment, initiate rule-based behavior. For instance, a tone produced by an electronic scale when a preprogrammed weight is reached acts as a sign to stop pouring milk. Symbols, which are meaningful formal structures that represent the functional properties of the environment, prompt knowledge-based behavior. As a case in point, a formula for converting a liter of milk into cups of milk may be stored as symbols in a mental model, thereby allowing one to calculate how many one-liter cartons are necessary for a recipe that requires eight cups.

A number of factors can influence whether information in the environment is interpreted as signals, signs, or symbols and, as a result, the level of cognitive control that is activated in performing a task. One factor is the *form* in which information is presented to workers. Other factors include the level of expertise of workers, the degree to which they reflect on their performance, and the nature of the task demands. These factors have implications for design.

Implications for design also arise from the observation that workers tend to execute lower levels of cognitive control more quickly, effectively, and effortlessly than higher levels. (Skill-based, rule-based, and knowledge-based behaviors represent progressively higher levels of control.) In addition, workers prefer performing tasks at lower levels of control, even when they are not explicitly supported to do so. This makes it clear that interfaces should present information in a form that does not force control to a level higher than that required by the task demands. Higher levels of control, though, may still be required because, as already noted, the level that is triggered in performing a task is based not only on the form of information presentation but also on a number of other factors. Clearly, then, interfaces need to support the competencies required at all three levels of cognitive control.

Summary

In this appendix, the key concepts of four of the dimensions of cognitive work analysis—those that were not the main subject of this book—were described briefly and illustrated with examples from a home. This appendix showed that control task analysis is concerned with establishing what needs to be done in a system, whereas strategies analysis involves examining how that activity can be done. Social organization and cooperation analysis deals with such important questions as who can carry out the work of the system or, more specifically, how this work can be allocated, distributed, and coordinated. Last, worker competencies analysis identifies the competencies required by workers to fulfill the system's work demands. This appendix served to remind readers that, beyond work domain analysis, the other dimensions of cognitive work analysis have significant parts to play in the design of effective complex sociotechnical systems.

References

Ahlstrom, U. (2005). Work domain analysis for air traffic controller weather displays. *Journal of Safety Research, 36,* 159–169.

Albrechtsen, H., and Pejtersen, A. M. (2003). Cognitive work analysis and work centered design of classification schemes. *Knowledge Organization, 30*(3–4), 213–227.

Amelink, M. H. J., Mulder, M., van Paassen, M. M., and Flach, J. (2005). Theoretical foundations for a total energy-based perspective flight-path display. *International Journal of Aviation Psychology, 15*(3), 205–231.

Baker, C. (2006). *Evaluation and refinement of a methodological approach for work domain analysis.* Unpublished honours thesis, RMIT University, Melbourne, VIC, Australia.

Baker, C., Naikar, N., and Neerincx, M. (2008). Engineering planetary exploration systems: Integrating novel technologies and the human element using work domain analysis. *Proceedings of the 59th International Astronautical Congress 2008: Symposium on Stepping Stones to the Future: Strategies, Architectures, Concepts and Technologies.* Glasgow, UK: International Astronautical Federation.

Benda, P. J., and Sanderson, P. M. (1998). Towards a dynamic model of adaptation to technological change. In P. Calder and B. Thomas (Eds.), *Proceedings of the Australasian Computer Human Interaction Conference (OzCHI98)* (pp. 244–251). Los Alamitos, CA: IEEE Computer Society Press.

Benda, P. J., and Sanderson, P. M. (1999). New technology and work practice: Modelling change with cognitive work analysis. In M. A. Sasse and C. Johnson (Eds.), *Proceedings of the Seventh IFIP TC13 International Conference on Human-Computer Interaction (INTERACT99)* (pp. 566–573). Amsterdam: IOS Press.

Bisantz, A. M., Roth, E., Brickman, B., Gosbee, L. L., Hettinger, L., and McKinney, J. (2003). Integrating cognitive analyses in a large-scale system design process. *International Journal of Human-Computer Studies, 58,* 177–206.

Bisantz, A. M., and Vicente, K. J. (1994). Making the abstraction hierarchy concrete. *International Journal of Human-Computer Studies, 40,* 83–117.

Borst, C., Suijkerbuijk, H. C. H., Mulder, M., and van Paassen, M. M. (2006). Ecological interface design for terrain awareness. *International Journal of Aviation Psychology, 16*(4), 375–400.

Bucciarelli, L. L. (1988). An ethnographic perspective on engineering design. *Design Studies, 9*(3), 159–168.

Burns, C. M. (1998). *The effects of spatial and temporal proximity of means-end information in ecological display design for an industrial simulation.* Unpublished doctoral dissertation, University of Toronto, Ontario, Canada.

Burns, C. M. (2000). Putting it all together: Improving display integration in ecological displays. *Human Factors, 42*(2), 226–241.

Burns, C. M., Bisantz, A. M., and Roth, E. M. (2004). Lessons from a comparison of work domain models: Representational choices and their implications. *Human Factors, 46*(4), 711–727.

Burns, C. M., Bryant, D. J., and Chalmers, B. A. (2001). Scenario mapping with work domain analysis. *Proceedings of the Human Factors and Ergonomics Society 45th Annual Meeting* (pp. 424–428). Santa Monica, CA: HFES.

Burns, C. M., Bryant, D. J., and Chalmers, B. A. (2005). Boundary, purpose, and values in work domain models: Models of naval command and control. *IEEE Transactions on Systems, Man, and Cybernetics—Part A: Systems and Humans, 35*(5), 603–616.

Burns, C. M., and Hajdukiewicz, J. R. (2004). *Ecological interface design.* Boca Raton, FL: CRC Press.

Burns, C. M., Kuo, J., and Ng, S. (2003). Ecological interface design: A new approach for visualizing network management. *Computer Networks, 43,* 369–388.

Burns, C. M., and Proulx, P. (2002). Influencing social problems with interface design. *Ergonomics in Design, 10*(4), 12–16.

Burns, C. M., and Vicente, K. J. (1995). A framework for describing and understanding interdisciplinary interactions in design. *Proceedings of the First Conference on Designing Interactive Systems: Processes, Practices, Methods, & Techniques (DIS '95)* (pp. 97–103). New York: ACM; doi: 10.1145/225434.225445.

Burns, C. M., and Vicente, K. J. (2000). A participant-observer study of ergonomics in engineering design: How constraints drive design process. *Applied Ergonomics, 31*, 73–82.

Burns, C. M., and Vicente, K. J. (2001). Model-based approaches for analyzing cognitive work: A comparison of abstraction hierarchy, multilevel flow modeling, and decision ladder modeling. *International Journal of Cognitive Ergonomics, 5*(3), 357–366.

Charlton, S. G., and O'Brien, T. G. (1996). The role of human factors testing and evaluation in systems development. In T. G. O'Brien and S. G. Charlton (Eds.), *Handbook of human factors testing and evaluation* (pp. 13–26). Mahwah, NJ: Erlbaum.

Childs, J. M., and Bell, H. H. (2002). Training systems evaluation. In S. G. Charlton and T. G. O'Brien (Eds.), *Handbook of human factors testing and evaluation* (2nd ed., pp. 473–509). Mahwah, NJ: Erlbaum.

Cooke, N. J. (1999). Knowledge elicitation. In F. T. Durso, R. S. Nickerson, R. W. Schvaneveldt, S. T. Dumais, D. S. Lindsay, and M. T. H. Chi (Eds.), *Handbook of applied cognition* (pp. 479–509). Chichester, UK: Wiley.

Czaja, S. J. (1997). Systems design and evaluation. In G. Salvendy (Ed.), *Handbook of human factors and ergonomics* (2nd ed., pp. 17–40). New York: Wiley.

Dainoff, M. J., and Mark, L. S. (1995). Use of a means-end abstraction hierarchy to conceptualize the ergonomic design of workplaces. In J. Flach, P. Hancock, J. Caird, and K. Vicente (Eds.), *Global perspectives on the ecology of human-machine systems* (pp. 273–292). Hillsdale, NJ: Erlbaum.

Davis, L. E., and Wacker, G. L. (1982). Job design. In G. Salvendy (Ed.), *Handbook of industrial engineering* (pp. 2.5.1–2.5.31). New York: Wiley.

Davis, L. E., and Wacker, G. L. (1987). Job design. In G. Salvendy (Ed.), *Handbook of human factors* (pp. 431–452). New York: Wiley.

de Groot, A. D. (1965). *Thought and choice in chess*. The Hague, Netherlands: Mouton.

Department of the Air Force. (1993). *Instructional system development* (AF Manual 36-2234). Washington, DC: Headquarters U.S. Air Force.

Department of Defence. (1995). *The capital equipment procurement manual* (CEPMAN 1). Canberra, ACT, Australia: Director Defence Publishing Service.

Department of Defence. (1999). *The defence procurement policy manual*. Canberra, ACT, Australia: Defence Publishing Service.

Dick, W., and Carey, L. (1990). *The systematic design of instruction* (3rd ed.). Tallahassee, FL: HarperCollins.

Dick, W., Carey, L., and Carey, J. O. (2005). *The systematic design of instruction* (6th ed.). Boston: Allyn & Bacon.

Dieterly, D. L. (1988). Team performance requirements. In S. Gael (Ed.), *The job analysis handbook for business, industry, and government* (pp. 766–777). New York: Wiley.

Dinadis, N., and Vicente, K. J. (1996). Ecological interface design for a power plant feedwater subsystem. *IEEE Transactions on Nuclear Science, 43*(1), 266–277.

Dinadis, N., and Vicente, K. J. (1999). Designing functional visualizations for aircraft systems status displays. *International Journal of Aviation Psychology, 9*(3), 241–269.

Dorneich, M. C. (2002). A system design framework-driven implementation of a learning collaboratory. *IEEE Transactions on Systems, Man, and Cybernetics—Part A: Systems and Humans, 32*(2), 200–213.

Dubrovsky, V., and Piscoppel, A. (1991). Toward a framework for structured job-collaboration design. *Proceedings of the Human Factors Society 35th Annual Meeting* (pp. 959–963). Santa Monica, CA: Human Factors Society.

Duez, P., and Vicente, K. J. (2005). Ecological interface design and computer network management: The effects of network size and fault frequency. *International Journal of Human-Computer Studies*, *63*, 565–586.

Duncker, K. (1945). On problem solving. Trans. L. S. Lees. *Psychological Monographs*, *58* (5, Whole No. 270).

Elix, B., and Naikar, N. (2008a). Designing safe and effective future systems: A new approach for modelling decisions in future systems with cognitive work analysis. *Proceedings of the Eighth International Symposium of the Australian Aviation Psychology Association*. Sydney, NSW, Australia: Australian Aviation Psychology Association.

Elix, B., and Naikar, N. (2008b). Unpublished work domain analysis of a maritime surveillance system. Defence Science and Technology Organisation, Fishermans Bend, VIC, Australia.

Elix, B., and Naikar, N. (2008c). Unpublished work domain analysis of an uninhabited aerial system. Defence Science and Technology Organisation, Fishermans Bend, VIC, Australia.

Ellerbroek, J., Visser, M., van Dam, S. B. J., Mulder, M., and van Paassen, M. M. (2009). Towards an ecological four-dimensional self-separation assistance display. *Proceedings of the AIAA Modeling and Simulation Technologies Conference* (pp. 1287–1299). Reston, VA: American Institute of Aeronautics and Astronautics.

Fidel, R., and Pejtersen, A. M. (2004). From information behaviour research to the design of information systems: The cognitive work analysis framework. *Information Research*, *10*, no. 1, paper 210 (October). http://InformationR.net/ir/10–1/paper210.html.

Frijda, N. H., and de Groot, A. D. (Eds.). (1981). *Otto Selz: His contribution to psychology*. The Hague, Netherlands: Mouton.

Gabb, A. P., and Henderson, D. E. (1995). *A review of Navy's technical and operational tender evaluation practices* (DSTO-TR-0194). Salisbury, SA, Australia: DSTO Electronics and Surveillance Research Laboratory.

Gabb, A. P., and Henderson, D. E. (1996). *Technical and operational tender evaluations for complex military systems* (DSTO-TR-0303). Salisbury, SA, Australia: DSTO Electronics and Surveillance Research Laboratory.

Gagne, R. M., Wager, W. W., Golas, K. C., and Keller, J. M. (2005). *Principles of instructional design* (5th ed.). Belmont, CA: Thomson Wadsworth.

Hackman, J. R., and Oldham, G. R. (1980). *Work redesign*. Reading, MA: Addison-Wesley.

Hajdukiewicz, J. R. (1998). *Development of a structured approach for patient monitoring in the operating room*. Unpublished master's thesis, University of Toronto, Ontario, Canada.

Hajdukiewicz, J. R., Burns, C. M., Vicente, K. J., and Eggleston, R. G. (1999). Work domain analysis for intentional systems. *Proceedings of the Human Factors and Ergonomics Society 43rd Annual Meeting* (pp. 333–337). Santa Monica, CA: HFES.

Hajdukiewicz, J. R., Doyle, D. J., Milgram, P., Vicente, K. J., and Burns, C. M. (1998). A work domain analysis of patient monitoring in the operating room. *Proceedings of the Human Factors and Ergonomics Society 42nd Annual Meeting* (pp. 1038–1042). Santa Monica, CA: HFES.

Hajdukiewicz, J. R., Vicente, K. J., Doyle, D. J., Milgram, P., and Burns, C. M. (2001). Modeling a medical environment: An ontology for integrated medical informatics design. *International Journal of Medical Informatics*, *62*, 79–99.

Ham, D. H., Yoon, W. C., and Han, B. T. (2008). Experimental study on the effects of visualized functionally abstracted information on process control tasks. *Reliability Engineering and System Safety*, *93*, 254–270.

Hansen, J. P., Løvborg, L., and Rasmussen, J. (1991). Simulation of cognitive behaviour in computer games. *Proceedings of the Second MOHAWC Workshop*. Roskilde, Denmark: Risø National Laboratory.

Hays, R. T., and Singer, M. J. (1989). *Simulation fidelity in training system design: Bridging the gap between reality and training*. New York: Springer-Verlag.

Hoffman, R. R., Crandall, B., and Shadbolt, N. (1998). Use of the critical decision method to elicit expert knowledge: A case study in the methodology of cognitive task analysis. *Human Factors*, *40*(2), 254–276.

Hopcroft, R., and Naikar, N. (2005). Unpublished work domain analysis of the Joint Strike Fighter. Defence Science and Technology Organisation, Fishermans Bend, VIC, Australia.

Huey, B. M., and Wickens, C. D. (Eds.). (1993). *Workload transition: Implications for individual and team performance*. Washington, DC: National Academy Press.

Itoh, J., Sakuma, A., and Monta, K. (1995). An ecological interface for supervisory control of BWR nuclear power plants. *Control Engineering Practice, 3*(2), 231–239.

Itoh, J., Yoshimura, S., Ohtsuka, T., and Masuda, F. (1990). Cognitive task analysis of nuclear power plant operators for man-machine interface design. *Proceedings of the American Nuclear Society Topical Meeting on Advances in Human Factors Research on Man-Computer Interactions: Nuclear and Beyond* (pp. 96–102). La Grange Park, IL: American Nuclear Society.

Jamieson, G. A. (1998). *Ecological interface design for petrochemical processing applications.* Unpublished master's thesis, University of Toronto, Ontario, Canada.

Jamieson, G. A., and Vicente, K. J. (2001). Ecological interface design for petrochemical applications: Supporting operator adaptation, continuous learning, and distributed, collaborative work. *Computers & Chemical Engineering, 25,* 1055–1074.

Jenkins, D. P., Stanton, N. A., Salmon, P. M., Walker, G. H., and Young, M. S. (2008). Using cognitive work analysis to explore activity allocation within military domains. *Ergonomics, 51*(6), 798–815.

Jenkins, D. P., Stanton, N. A., Walker, G. H., Salmon, P. M., and Young, M. S. (2010). Using cognitive work analysis to explore system flexibility. *Theoretical Issues in Ergonomics Science, 11*(3), 136–150.

Jenkins, D. P., Stanton, N. A., Walker, G. H., and Young, M. S. (2007). A new approach to designing lateral collision warning systems. *International Journal of Vehicle Design, 45*(3), 379–396.

Kinsley, A. M., Sharit, J., and Vicente, K. J. (1994). Abstraction hierarchy representation of manufacturing: Towards ecological interfaces for advanced manufacturing systems. In P. T. Kidd and W. Karwowski (Eds.), *Advances in agile manufacturing* (pp. 297–300). Amsterdam: IOS Press.

Kirwan, B., and Ainsworth, L. K. (Eds.). (1992). *A guide to task analysis.* London: Taylor & Francis.

Klein, G. A., Calderwood, R., and MacGregor, D. (1989). Critical decision method for eliciting knowledge. *IEEE Transactions on Systems, Man, and Cybernetics, 19*(3), 462–472.

Klein, G., and Militello, L. (2001). Some guidelines for conducting a cognitive task analysis. In E. Salas (Ed.), *Advances in human performance and cognitive engineering research* (Vol. 1, pp. 163–199). New York: Elsevier Science.

Lambeth, S., and Naikar, N. (2008). Unpublished work domain analysis of an air power system. Defence Science and Technology Organisation, Fishermans Bend, VIC, Australia.

Lau, N., Jamieson, G. A., Skraaning, G., Jr., and Burns, C. M. (2008). Ecological interface design in the nuclear domain: An empirical evaluation of ecological displays for the secondary subsystems of a boiling water reactor plant simulator. *IEEE Transactions on Nuclear Science, 55*(6), 3597–3610.

Lehner, P. E. (1991). Towards a prescriptive theory of team design. *Proceedings of the IEEE International Conference on Systems, Man, and Cybernetics: Decision Aiding for Complex Systems* (pp. 2029–2034). New York: IEEE.

Leveson, N. G. (2000). Intent specifications: An approach to building human-centered specifications. *IEEE Transactions on Software Engineering, 26*(1), 15–35.

Lind, M. (2003). Making sense of the abstraction hierarchy in the power plant domain. *Cognition, Technology & Work, 5,* 67–81.

Linegang, M. P., and Lintern, G. (2003). Multifunction displays for optimum manning: Towards functional integration and cross-functional awareness. *Proceedings of the Human Factors and Ergonomics Society 47th Annual Meeting* (pp. 1923–1927). Santa Monica, CA: HFES.

Lintern, G. (2006). A functional workspace for military analysis of insurgent operations. *International Journal of Industrial Ergonomics, 36*(5), 409–422.

Lintern, G., and Miller, C. (2003). Identification of cognitive demands for new systems. *Proceedings of the Human Factors and Ergonomics Society 47th Annual Meeting* (pp. 1933–1937). Santa Monica, CA: HFES.

Malone, T. B. (1996). Human factors test support documentation. In T. G. O'Brien and S. G. Charlton (Eds.), *Handbook of human factors testing and evaluation* (pp. 101–116). Mahwah, NJ: Erlbaum.

Mazaeva, N., and Bisantz, A. M. (2007). On the representation of automation using a work domain analysis. *Theoretical Issues in Ergonomics Science*, 8(6), 509–530.

Medsker, G. J., and Campion, M. A. (1997). Job and team design. In G. Salvendy (Ed.), *Handbook of human factors and ergonomics* (2nd ed., pp. 450–489). New York: Wiley.

Meister, D. (1996). Human factors test and evaluation in the twenty-first century. In T. G. O'Brien and S. G. Charlton (Eds.), *Handbook of human factors testing and evaluation* (pp. 313–322). Mahwah, NJ: Erlbaum.

Miller, A. (2004). A work domain analysis framework for modelling intensive care unit patients. *Cognition, Technology & Work*, 6(4), 207–222.

Miller, A., and Sanderson, P. (2000). Modeling "deranged" physiological systems for ICU information system design. *Proceedings of the Joint 14th Triennial Congress of the International Ergonomics Association/44th Annual Meeting of the Human Factors and Ergonomics Society (HFES/IEA 2000), 4* (pp. 245–248). Santa Monica, CA: HFES.

Miller, C. A., and Vicente, K. J. (1998). *Abstraction decomposition space analysis for NOVA's E1 acetylene hydrogenation reactor* (CEL 98–09). Toronto, Ontario, Canada: Cognitive Engineering Laboratory.

Miller, R. B. (1953). *A method for man-machine task analysis* (WADC Tech. Rep. No. 53-137). Wright-Patterson Air Force Base, OH: Wright Air Development Center.

Moray, N., Sanderson, P. M., and Vicente, K. J. (1992). Cognitive task analysis of a complex work domain: A case study. *Reliability Engineering and System Safety*, 36, 207–216.

Morel, G., and Chauvin, C. (2006). A socio-technical approach of risk management applied to collisions involving fishing vessels. *Safety Science*, 44, 599–619.

Mundel, M. E. (1985). *Motion and time study: Improving productivity* (6th ed.). Englewood Cliffs, NJ: Prentice Hall.

Nadimian, R. M., Griffiths, S., and Burns, C. M. (2002). Ecological interface design in aviation domains: Work domain analysis and instrumentation availability of the Harvard aircraft. *Proceedings of the Human Factors and Ergonomics Society 46th Annual Meeting* (pp. 116–120). Santa Monica, CA: HFES.

Naikar, N. (2006a). A comparison of the decision ladder template and the recognition-primed decision model. Paper prepared for the International Workshop on Intelligent Decision Support Systems: Retrospect and Prospects, 29 August–2 September 2005, Siena, Italy.

Naikar, N. (2006b). An examination of the key concepts of the five phases of cognitive work analysis with examples from a familiar system. *Proceedings of the Human Factors and Ergonomics Society 50th Annual Meeting* (pp. 447–451). Santa Monica, CA: HFES.

Naikar, N. (2006c). Beyond interface design: Further applications of cognitive work analysis. *International Journal of Industrial Ergonomics*, 36, 423–438.

Naikar, N. (2009). Beyond the design of ecological interfaces: Applications of work domain analysis and control task analysis to the evaluation of design proposals, team design, and training. In A. M. Bisantz and C. M. Burns (Eds.), *Applications of cognitive work analysis* (pp. 69–94). Boca Raton, FL: CRC Press.

Naikar, N. (2010). *A comparison of the decision ladder template and the recognition-primed decision model* (DSTO-TR-2397). Fishermans Bend, VIC, Australia: Defence Science and Technology Organisation.

Naikar, N., Hopcroft, R., and Moylan, A. (2005). *Work domain analysis: Theoretical concepts and methodology* (DSTO-TR-1665). Fishermans Bend, VIC, Australia: Defence Science and Technology Organisation.

Naikar, N., Moylan, A., and Pearce, B. (2006). Analysing activity in complex systems with cognitive work analysis: Concepts, guidelines, and case study for control task analysis. *Theoretical Issues in Ergonomics Science*, 7(4), 371–394.

Naikar, N., Pearce, B., Drumm, D., and Sanderson, P. M. (2003). Designing teams for first-of-a-kind, complex systems using the initial phases of cognitive work analysis: Case study. *Human Factors*, 45(2), 202–217.

Naikar, N., and Sanderson, P. M. (1999). Work domain analysis for training-system definition and acquisition. *International Journal of Aviation Psychology*, 9(3), 271–290.

Naikar, N., and Sanderson, P. M. (2001). Evaluating design proposals for complex systems with work domain analysis. *Human Factors, 43*(4), 529–542.

Newell, A., and Simon, H. A. (1972). *Human problem solving*. Englewood Cliffs, NJ: Prentice Hall.

Niebel, B. W. (1988). *Motion and time study* (8th ed.). Homewood, IL: Irwin.

O'Brien, T. G. (1996). Preparing human factors test plans and reports. In T. G. O'Brien and S. G. Charlton (Eds.), *Handbook of human factors testing and evaluation* (pp. 117–134). Mahwah, NJ: Erlbaum.

Pejtersen, A. M., Sonnenwald, D. H., Buur, J., Govindaraj, T., and Vicente, K. (1997). The design explorer project: Using a cognitive framework to support knowledge exploration. *Journal of Engineering Design, 8*(3), 289–301.

Perrow, C. (1984). *Normal accidents: Living with high-risk technologies*. New York: Basic Books.

Rasmussen, J. (1969). *Man-machine communication in the light of accident records* (Report No. S-1-69). Roskilde, Denmark: Danish Atomic Energy Commission, Research Establishment Risø.

Rasmussen, J. (1974). *The human data processor as a system component: Bits and pieces of a model* (Risø-M-1722). Roskilde, Denmark: Danish Atomic Energy Commission, Research Establishment Risø.

Rasmussen, J. (1976). Outlines of a hybrid model of the process plant operator. In T. B. Sheridan and G. Johannsen (Eds.), *Monitoring behavior and supervisory control* (pp. 371–383). New York: Plenum Press.

Rasmussen, J. (1979). *On the structure of knowledge—A morphology of mental models in a man-machine system context* (Risø-M-2192). Roskilde, Denmark: Risø National Laboratory.

Rasmussen, J. (1981). Models of mental strategies in process plant diagnosis. In J. Rasmussen and W. B. Rouse (Eds.), *Human detection and diagnosis of system failures* (pp. 241–258). New York: Plenum Press.

Rasmussen, J. (1983). Skills, rules, and knowledge; signals, signs, and symbols, and other distinctions in human performance models. *IEEE Transactions on Systems, Man, and Cybernetics, 13*(3), 257–266.

Rasmussen, J. (1985). The role of hierarchical knowledge representation in decisionmaking and system management. *IEEE Transactions on Systems, Man, and Cybernetics, 15*(2), 234–243.

Rasmussen, J. (1986). *Information processing and human-machine interaction: An approach to cognitive engineering*. New York: North-Holland.

Rasmussen, J. (1988). A cognitive engineering approach to the modeling of decision making and its organization in: Process control, emergency management, CAD/CAM, office systems, and library systems. In W. B. Rouse (Ed.), *Advances in man-machine systems research, 4* (pp. 165–243). Greenwich, CT: JAI Press.

Rasmussen, J. (1990). A model for the design of computer integrated manufacturing systems: Identification of information requirements of decision makers. *International Journal of Industrial Ergonomics, 5*, 5–16.

Rasmussen, J. (1996). *Modeling socio-technical systems: A risk managing perspective*. Working Paper, Risk Center, University of Karlstad, Sweden.

Rasmussen, J. (1997). Merging paradigms: Decision making, management, and cognitive control. In R. Flin, E. Salas, M. Strub, and L. Martin (Eds.), *Decision making under stress: Emerging themes and applications* (pp. 67–81). Aldershot, UK: Ashgate.

Rasmussen, J. (1998). *Ecological interface design for complex systems: An example: SEAD–UAV systems* (AFRL-HE-WP-TR-1999–0011). Wright-Patterson Air Force Base, OH: Air Force Research Laboratory, Human Effectiveness Directorate.

Rasmussen, J. (1999). Ecological interface design for reliable human-machine systems. *International Journal of Aviation Psychology, 9*(3), 203–223.

Rasmussen, J., and Goodstein, L. P. (1987). Decision support in supervisory control of high-risk industrial systems. *Automatica, 23*(5), 663–671.

Rasmussen, J., and Jensen, A. (1974). Mental procedures in real-life tasks: A case study of electronic trouble shooting. *Ergonomics, 17*(3), 293–307.

Rasmussen, J., Pedersen, O. M., and Grønberg, C. D. (1987). *Evaluation of the use of advanced informa- tion technology (expert systems) for data base system development and emergency management in non- nuclear industries* (Risø-M-2639). Roskilde, Denmark: Risø National Laboratory.

Rasmussen, J., and Pejtersen, A. M. (1995). Virtual ecology of work. In J. Flach, P. Hancock, J. Caird, and K. Vicente (Eds.), *Global perspectives on the ecology of human-machine systems* (pp. 121–156). Hillsdale, NJ: Erlbaum.

Rasmussen, J., Pejtersen, A. M., and Goodstein, L. P. (1994). *Cognitive systems engineering*. New York: Wiley.

Rasmussen, J., Pejtersen, A. M., and Schmidt, K. (1990). *Taxonomy for cognitive work analysis* (Risø-M-2871). Roskilde, Denmark: Risø National Laboratory.

Rasmussen, J., and Vicente, K. J. (1989). Coping with human errors through system design: Implications for ecological interface design. *International Journal of Man-Machine Studies, 31*(5), 517–534.

Reason, J. (1990). *Human error*. Cambridge, UK: Cambridge University Press.

Reising, D. V. C. (1999). *The impact of instrumentation location and reliability on the performance of opera- tors using an ecological interface for process control*. Unpublished doctoral dissertation, University of Illinois at Urbana-Champaign.

Reising, D. V. C. (2000). The abstraction hierarchy and its extension beyond process control. *Proceedings of the Joint 14th Triennial Congress of the International Ergonomics Association/44th Annual Meeting of the Human Factors and Ergonomics Society (IEA/HFES 2000), 1* (pp. 194–196). Santa Monica, CA: HFES.

Reising, D. V. C., and Sanderson, P. M. (2002a). Ecological interface design for Pasteurizer II: A pro- cess description of semantic mapping. *Human Factors, 44*(2), 222–247.

Reising, D. V. C., and Sanderson, P. M. (2002b). Work domain analysis and sensors II: Pasteurizer II case study. *International Journal of Human-Computer Studies, 56*(6), 597–637.

Roth, E. M., and Mumaw, R. J. (1995). Using cognitive task analysis to define human interface require- ments for first-of-a-kind systems. *Proceedings of the Human Factors and Ergonomics Society 39th Annual Meeting* (pp. 520–524). Santa Monica, CA: HFES.

Schraagen, J. M., Chipman, S. F., and Shalin, V. L. (Eds.). (2000). *Cognitive task analysis*. Mahwah, NJ: Erlbaum.

Seamster, T. L., Redding, R. E., and Kaempf, G. L. (1997). *Applied cognitive task analysis in aviation*. Aldershot, UK: Ashgate.

Sharp, T. D. (1996). *Progress towards a development methodology for decision support systems for use in time-critical, highly uncertain, and complex environments*. Unpublished doctoral dissertation, University of Cincinnati, OH.

Shepherd, A. (2001). *Hierarchical task analysis*. London: Taylor & Francis.

Simon, H. A. (1981). *The sciences of the artificial* (2nd ed.). Cambridge, MA: MIT Press.

Simons, K. J., Dainoff, M. J., and Mark, L. S. (2007). The work process of research librarians, elicited via the abstraction–decomposition space. *Advances in Library Administration and Organization, 24*, 191–230.

Stoner, H. A., Wiese, E. E., and Lee, J. D. (2003). Applying ecological interface design to the driving domain: The results of an abstraction hierarchy analysis. *Proceedings of the Human Factors and Ergonomics Society 47th Annual Meeting* (pp. 444–448). Santa Monica, CA: HFES.

Sundstrom, E., De Meuse, K. P., and Futrell, D. (1990). Work teams: Applications and effectiveness. *American Psychologist, 45*(2), 120–123.

Taylor, F. W. (1911). *The principles of scientific management*. New York: Harper & Bros.

Thompson, L. K., Hickson, J. C. L., and Burns, C. M. (2003). A work domain analysis for diabetes management. *Proceedings of the Human Factors and Ergonomics Society 47th Annual Meeting* (pp. 1516–1520). Santa Monica, CA: HFES.

Torenvliet, G. L., Jamieson, G. A., and Chow, R. (2008). Object worlds in work domain analysis: A model of naval damage control. *IEEE Transactions on Systems, Man, and Cybernetics—Part A: Systems and Humans, 38*(5), 1030–1040.

Treadwell, A., and Naikar, N. (2009). Unpublished work domain analysis of an air combat system. Defence Science and Technology Organisation, Fishermans Bend, VIC, Australia.

Van Dam, S. B. J., Mulder, M., and van Paassen, M. M. (2008). Ecological interface design of a tactical airborne separation assistance tool. *IEEE Transactions on Systems, Man, and Cybernetics—Part A: Systems and Humans, 38*(6), 1221–1233.

Vicente, K. J. (1999). *Cognitive work analysis: Toward safe, productive, and healthy computer-based work.* Mahwah, NJ: Erlbaum.

Vicente, K. J. (2002). Ecological interface design: Progress and challenges. *Human Factors, 44*(1), 62–78.

Vicente, K. J., Christoffersen, K., and Pereklita, A. (1995). Supporting operator problem solving through ecological interface design. *IEEE Transactions on Systems, Man, and Cybernetics, 25*(4), 529–545.

Vicente, K. J., and Rasmussen, J. (1990). The ecology of human-machine systems II: Mediating "direct perception" in complex work domains. *Ecological Psychology, 2*(3), 207–249.

Vicente, K. J., and Rasmussen, J. (1992). Ecological interface design: Theoretical foundations. *IEEE Transactions on Systems, Man, and Cybernetics, 22*(4), 589–606.

Vicente, K. J., and Wang, J. H. (1998). An ecological theory of expertise effects in memory recall. *Psychological Review, 105*(1), 33–57.

Wallace, P., Walkden, K., and Lintern, G. (1999). An analysis of F/A-18A pilot training tasks. *Proceedings of the Fourth International SimTect Conference* (pp. 15–19). Melbourne, VIC, Australia: SimTect 99 Organising and Technical Committee.

Watson, M., and Sanderson, P. (1998). Work domain analysis for the evaluation of human interaction with anaesthesia alarm systems. In P. Calder and B. Thomas (Eds.), *Proceedings of the Australasian Computer Human Interaction Conference (OzCHI98)* (pp. 228–235). Los Alamitos, CA: IEEE Computer Society Press.

Watson, M. O., and Sanderson, P. M. (2007). Designing for attention with sound: Challenges and extensions to ecological interface design. *Human Factors, 49*(2), 331–346.

Whitefield, A., Wilson, F., and Dowell, J. (1991). A framework for human factors evaluation. *Behaviour & Information Technology, 10*(1), 65–79.

Xu, W. (2007). Identifying problems and generating recommendations for enhancing complex systems: Applying the abstraction hierarchy framework as an analytical tool. *Human Factors, 49*(6), 975–994.

Xu, W., Dainoff, M. J., and Mark, L. S. (1999). Facilitate complex search tasks in hypertext by externalizing functional properties of a work domain. *International Journal of Human-Computer Interaction, 11*(3), 201–229.

Yeung, J., and Naikar, N. (2010). Unpublished work domain analysis of an air power system. Defence Science and Technology Organisation, Fishermans Bend, VIC, Australia.

Zsambok, C. E., and Klein, G. (Eds.). (1997). *Naturalistic decision making.* Mahwah, NJ: Erlbaum.

Author Index

Bold typeface indicates the lead author.

A

Ahlstrom (2005), 58t
Ainsworth, **Kirwan** and (1992), *see* **Kirwan** and Ainsworth (1992)
Albrechtsen and Pejtersen (2003), 58t
Amelink, Mulder, van Paassen, and Flach (2005), 58t

B

Baker (2006)
    analytic themes, work domain analysis, 154
    decomposition dimension, 192
    indeterminate set, abstraction dimension, 188
Baker, Naikar, and Neerincx (2008)
    analytic themes, work domain analysis, 154
    decomposition dimension, 192
    indeterminate set, abstraction dimension, 188
Bell, **Childs** and (2002), 260
Benda and Sanderson (1998)
    engineering design systems, work domain models, 58t
    multiple stakeholders' perspectives, 119, 120f, 123, 125
Benda and Sanderson (1999), 127
Bisantz, **Mazaeva** and (2007), *see* **Mazaeva** and Bisantz (2007)
Bisantz, Roth, Brickman, Gosbee, Hettinger, and McKinney (2003)
    focus system's position, causal–intentional continuum, 170
    interviews, 177, 178
    military systems, work domain models, 58t
    multiple problem facets, 133, 134f
Bisantz, and Roth, **Burns** (2004), *see* **Burns**, Bisantz, and Roth (2004)
Bisantz and Vicente (1994)
    causal *versus* intentional systems, 40, 42f
    focus system's position, causal–intentional continuum, 170
    graphical abstraction–decomposition space, 45

    number, types, and labels of levels of decomposition, 78
    process control systems, work domain models, 58t
    system decomposition, 89
    topological relations, 111, 112f
Borst, Suijkerbuijk, Mulder, and van Paassen (2006), 58t
Brickman, Gosbee, Hettinger, and McKinney, **Bisantz**, Roth (2003), *see* **Bisantz**, Roth, Brickman, Gosbee, Hettinger, and McKinney (2003)
Bryant, and Chalmers, **Burns** (2001), 203, 204
Bryant, and Chalmers, **Burns** (2005), *see* **Burns**, Bryant, and Chalmers (2005)
Bucciarelli (1988), 117, 118
Burns (1998), 111
Burns (2000)
    number, types, and labels of levels of abstraction, 59
    number, types, and labels of levels of decomposition, 78
    process control systems, work domain models, 58t
    standard set, abstraction dimension, 183
Burns, Bisantz, and Roth (2004)
    control systems, inclusion in work domain model, 144
    multiple problem facets, 133
Burns, Bryant, and Chalmers (2001), 203, 204
Burns, Bryant, and Chalmers (2005)
    boundaries of the analysis, 159
    causal *versus* intentional systems, 38
    focus system's position, causal–intentional continuum, 170
    military systems, work domain models, 58t
    multiple problem facets, 127, 129, 130f, 131f, 132f, 133
    multiple stakeholders' perspectives, object worlds, 125
    number, types, and labels of levels of abstraction, 59
    scenario mapping, 203, 204
    subject matter experts, review with, 203
Burns and Hajdukiewicz (2004)
    abstraction–decomposition space, validity of model of focus system, 201

Subject Index

A

Abstract function level, 111, 114

Abstraction
  content of abstraction–decomposition space
    indeterminate abstraction levels, 184–188
    standard abstraction levels, 181–184
  descriptions of levels
    functional purposes, 67–69
    object-related processes, 73–74
    overview, 64–67
    physical objects, 74–75
    purpose-related functions, 71–72
    value and priority measures, 70–71
  dimension, 24, 28–29, 30, 31, 36, 37, 99, 180
  number, types, and labels of levels, 25, 28, 51–64
  physical concepts, 28, 36, 37, 40
  purposive concepts, 28, 36, 37, 40
  reasoning at different levels, 36

Abstraction–decomposition space, *see also* Abstraction hierarchy
  abstraction dimension, 28–29
  constraints, 29–30
  decomposition dimension, 29
  description, 22–23, 24–25, 33–37, 180
  formats, work domain model, 44–46
  home model, 25–28
  multiple models and decomposition, 135
  overview, 24–25
  part–whole relations, 24, 29
  relevance to human reasoning, 35–37
  structural means–ends relations, 24, 30–33
  tool for modeling work domain, 7, 33–35

Abstraction–decomposition space, validity
  hints, 204
  overview, 201–202
  scenario mapping, 203–204
  subject matter experts, 202–203

Abstraction hierarchy, *see also* Abstraction–decomposition space
  decomposition, 89, 92–97
  description, 25
  formats, work domain model, 45–46
  modeling tool, 7, 25
  purpose-oriented reasoning, 37

Acetylene hydrogenation reactor example
  decomposition, 78, 80

  part–whole relations, 107
  scenario mapping, 204
  system decomposition, 89

Action means–ends relations, 31, 37, 99–100, 139–141, 142, 144, 210

Action-relevance, *xvii*, 19

Activity, *see also* Control task analysis; Strategies analysis
  analysis, 268
  control systems, *xvii*, 143–146
  inclusion or exclusion in work domain model, 139–141
  noun or verb usage, 141–142
  order of dimensions of cognitive work analysis, 12–13
  overlap, work domain analysis and control task analysis, 146, 148

Actors
  behavior representation, 19–23, 33–34
  boundaries of analysis, 123, 125, 127, 159, 164
  constraints, from complex sociotechnical systems, 5
  control systems, inclusion in work domain model, 144
  coping with complexity, 37
  defined, *xvii*
  focus system's position on causal–intentional continuum, 38–43, 167
  instantiations of means–ends relations, 107
  multiple stakeholders' perspectives, 123, 125, 127, 164
  subjective task formulation, 271
  tool for modeling work domain, 33–35

Actors' perspective, boundaries of analysis, 125, 127, 159, 160, 162, 170, 171, 185, 205, 213, 232, 253, *see also* Multiple stakeholders' perspectives

Adaptation, designing for, 5, 15–17

AEW&C, *see* Airborne Early Warning and Control system case study

Affordance
  activities, work domain model, 140
  constraints, 4–5, 23
  noun or verb usage, 235
  overlap, work domain analysis and control task analysis, 148
  physical objects, 73, 74